THIRD EDITION

HINTS FOR THE HIGHLY EFFECTIVE INSTRUCTOR

SURVIVAL SKILLS FOR THE TECHNICAL TRAINER

AMERICAN TECHNICAL PUBLISHERS
ORLAND PARK, ILLINOIS 60467-5756

W. R. Miller
M. F. Miller

American Technical Publishers, Inc., Editorial Staff

Editor in Chief:
Jonathan F. Gosse
Vice President—Production:
Peter A. Zurlis
Art Manager:
James M. Clarke
Copy Editors:
Catherine A. Mini
Talia J. Lambarki
Cover Design:
James M. Clarke
Illustration/Layout:
Nicole D. Bigos

3 4 5 6 7 8 9 – 11 – 9 8 7 6 5 4 3 2 1

Printed in the United States of America

ISBN 978-0-8269-4146-6

This book is printed on recycled paper.

Table of Contents

HINTS FOR THE HIGHLY EFFECTIVE INSTRUCTOR

Introduction

HINTS FOR THE HIGHLY EFFECTIVE INSTRUCTOR

An instructional program is most effective when the instructor understands and addresses the complexities of the teaching-learning process. *Hints for the Highly Effective Instructor: Survival Skills for the Technical Trainer,* 3rd Edition, includes strategies and techniques for making instruction and learning effective and efficient. Designed for both new and experienced instructors, the information presented is organized for quick access and easy implementation. Topics such as instructional methods and resources, organization of the learning environment, and professional development of the instructor are included in the first nine chapters. The last chapter includes questions that can be used as a self-assessment to evaluate instructional strategies and techniques.

The 3rd Edition also addresses instructional challenges resulting from learner differences between those who grew up with digital technology (digital natives) and those who adopted digital technology later in life (digital immigrants). In addition, information on instructional technology and on-line course delivery has been expanded.

This valuable, concise reference can be used as a guide for implementing new instructional programs or for affirming the attributes of existing successful instructional programs.

The Publisher

About the Authors

HINTS FOR THE HIGHLY EFFECTIVE INSTRUCTOR

Dr. Wilbur R. Miller has taught at both the secondary and postsecondary levels and served as dean of the College of Education at the University of Missouri-Columbia. He has received numerous awards and citations, including the Distinguished Service Award from the National Center for Research in Career and Technical Education and the University of Missouri Alumni Faculty Award.

Dr. Miller has authored or co-authored eight books and many periodicals, monographs, and bulletins. He serves as a consultant to schools and industry in the U.S. and other countries. Dr. Miller has served and Assistant Vice-President, Associate Vice-President, and Vice-President for Development at Auburn University.

Dr. Marie F. Miller received a PhD in technical education from the University of Missouri-Columbia. She has received many awards and citations for her professional contributions, including the Distinguished Alumni Award from the University of Wisconsin-Stout and the Undergraduate Teaching Excellence Award from the Auburn University Alumni Association.

Dr. Miller has authored and co-authored four books and has written numerous articles and papers. She serves on the Board of Editors for the Journal of Technology Studies and as Test Center Coordinator for Alabama for the National Occupational Competency Testing Institute.

Dr. Miller holds the Endowed Mildred Cheshire Fraley Distinguished Professorship in the Department of Educational Foundations, Leadership, and Technology at Auburn University.

Introduction

An instructor's personal characteristics come through in the classroom setting. Attitudes, values, and communication style all influence the way learners respond. An instructor's professional behavior can serve to model behavior expected in the field. Also, instructors are responsible for certain administrative tasks. They are responsible as well for their own professional development, which can help them stay up-to-date and keep learners knowledgeable about necessary competencies and current issues.

Convey Enthusiasm for Teaching and Respect for Learners

Many variables contribute to an instructor's effectiveness. Personal attributes of instructors such as intelligence, knowledge of subject matter, and background are important in the teaching-learning process. Effective teaching involves a complex set of variables, making the measurement of teaching effectiveness by using standardized instruments difficult and often unreliable. Effective instructors are those who can bring about changed behavior in the learners.

Instructors who are excited about teaching and learning convey a positive and accepting attitude to learners. They communicate the value of attaining certain skills and knowledge and hold high expectations for themselves and their learners. Also, effective instructors show consideration for different personalities and learning styles and preferences. Individual differences bring strengths, not weaknesses, to the classroom. Instructors should treat participants in specialized training and workshops with respect. In fact, an instructor's acceptance of and concern for the learner is one of the most important factors in fostering an individual's motivation to learn. *Special Populations in Career and Technical Education,* published by American Technical Publishers, Inc., includes current research on strategies for motivating individuals to learn.

Many participants in specialized instructional programs have had unpleasant experiences in school; it is critical that instructors of technical subjects show warmth and empathy to these learners. An instructor's professional appearance and courteous behavior toward learners and other school personnel communicate positive intentions and a willingness to help.

Practice Equitable and Fair Treatment

Learners, like instructors, have different personalities, expectations, interests, needs, and ability levels. Content should be presented so that individuals of all ability levels can understand. If an instructor shows favoritism and special consideration toward some learners, all learners are affected. Those who are favored may mislead instructors into thinking that they are doing a good job of teaching, while those who are not favored will quickly become discouraged.

From time to time, assignments may need to be varied. This is especially true when instruction is individualized; however, there should be adequate and challenging assignments and activities for everyone participating in a particular learning situation. *Instructors and Their Jobs,* published by American Technical Publishers, Inc., includes an excellent section on diversity in the classroom.

Practice Sound Judgment and Diplomacy

Instructors should not show anger or argue with learners during class; this demonstrates poor judgment on the part of the instructor. Occasionally, a learner may know or pretend to know more about a certain topic than the instructor. In such cases, the instructor should avoid a confrontation with the learner in the presence of other learners. The instructor has nothing to gain, but risks losing the respect of the class by insisting on being right. In such cases, instructors can avoid criticism by tactfully avoiding an open argument until the facts can be collected. Instructors should be diplomatic and know how to say, "I don't know, but I'll find out," or "Let's check that out."

Even when a learner disagrees with an instructor, it is very difficult to argue with diplomacy. In any case, whether the instructor is right or wrong, the documented facts should promptly be brought back to the class.

Cooperate with Colleagues and Practice Teamwork

Effective training programs operate in an efficient and businesslike manner. It is a mistake to fail to collaborate and cooperate with colleagues. Management can be effective only when there is teamwork. Cooperation is needed to conserve the time of the instructors, learners, and other personnel. Instructors who go above and beyond the minimal expectations of being a good team player are positive role models for their learners and garner the respect of their colleagues.

Instructors may be called upon or choose to team teach, dividing the instruction into areas of individual strengths or interests. For example, one instructor may conduct large group sessions and another may work with learners who are advanced or who may need additional instruction and support. Team teaching requires members of the team to make joint decisions about individual responsibilities, methods of instructional delivery, and use of time, materials, and other resources.

Be Punctual

Instructors who are in the classroom or laboratory organizing their work before the scheduled time for class to start may find the learners getting to class early or at least on schedule. Instructors who come to the class at precisely the scheduled time or a few minutes late may find that learners come in late. Learners may assume that if the instructor is not there on time, they don't need to be either.

The importance of a good example on the part of the instructor should be recognized. Being prompt shows initiative and helps develop the habit of getting to the job on time. In addition, being in the classroom before the scheduled time gives the instructor and the learners a few minutes to settle in and prepare. An added benefit is that the learners will have an opportunity to talk with the instructor and others in the class before formal instruction starts. This can be very important as learners may have questions about the class or laboratory assignment, other courses, or career opportunities.

Maintain a Professional Appearance

Participants in specialized training workshops and technical education programs often look to instructors for standards of behavior. An instructor may be one of only a few role models for these learners. Instructors who come to class clean and well-groomed project a positive image to the participants.

Appropriate dress may range from smart casual to work uniforms depending on the type of training and work to be done. Personal protective equipment may be an important and necessary part of dress. The manner in which an instructor dresses demonstrates to learners the standard for acceptable dress, and in turn, what is expected and accepted in the industry. Instructors who care about their own appearance model behavior that will stay with most of the learners throughout their careers.

Motivate and Encourage All Learners

When instructors provide support and encouragement, people learn more. If instructors are not supportive and encouraging, the learners may learn less, develop indifferent attitudes, doubt their self-worth, become frustrated, or leave the instructional program. All learners deserve the opportunity to experience success. Checking and evaluating assignments and providing immediate feedback are excellent means of encouragement. Learners want instructors to comment on the completeness, accuracy, neatness, and overall quality of their work in both the classroom and the laboratory.

Instructors should encourage learners who have done their best, even if the overall quality of the work is not at the desired level. When a learner is not performing up to par, the instructor should comment on areas in which the learner does well, and not emphasize the learner's mistakes. Such phrases as, "No, that is wrong," or "That is not right," are far more discouraging than phrases such as, "Let's try that again," or "That is a good start, how can you improve on it?" Generally, people will put forth high levels of effort if they feel it will help them meet their goals and satisfy their needs. (See *Handbook for College Teaching* distributed by American Technical Publishers, Inc., for more information on motivation and need satisfaction.)

When a learner makes a comment or offers an answer that is partly correct, the instructor should comment immediately on the part that is correct without criticizing or reprimanding the learner. Only through encouragement and reinforcement will most learners continue to improve. When a learner's work begins to improve, instructors should provide praise and recognize the improvement. However, complimenting a learner every time the learner makes a comment or answers a question correctly is not advisable. Excessive compliments may be embarrassing to the shy learner or may be misunderstood by the other learners as a show of favoritism. This is particularly true when the learner's contribution is rather ordinary. Each time a learner makes a comment or answers a question correctly, the instructor should acknowledge that the contribution is correct. This way all learners will know that the comment is correct.

Demonstrate Good Communication Skills

The ability to communicate effectively is key to getting a job, keeping a job, and being promoted in a job. All occupations continue to require more sophistication, not only in the technology of the field, but also in the way in which individuals communicate. It is not surprising that national initiatives and mandates require the integration of technical education and academic subjects. For this reason, effective communication skills should be of vital concern to instructors. For example, instructors who use correct English and proper pronunciation and spelling are modeling to learners that there is more to preparing for a job in an industrial or technical field than simply possessing the required technical knowledge and skills.

Instructors should take care to speak slowly and clearly so that learners can understand the vocabulary, especially the technical terminology, needed in their area of interest. Instructors should go beyond the definition of a term, if necessary, to illustrate the use of technical terminology in the subject area. Effective communication, both written and oral, requires that definitions, instructions, illustrations, assignments, and other explanations be presented in technically correct language that is easily understood. An instructor should always review and check all handout materials and tests to ensure that they do not contain misspelled words, poor punctuation, or inconsistencies.

After learners complete a specific training program, they should be prepared to incorporate their new knowledge and skills into their work. Learners who complete technical education programs will be expected to interact with customers and clients, prepare and read work orders, work as members of a team, and participate in discussions related to workplace policies and procedures. Successful learners are the best advertisement for a training program.

Instructors use multiple methods of communication, not only for instructional purposes, but also to interact with individual learners, other instructors, managers, or industry personnel. Email and voice mail are common forms of modern communication. These forms of communication are as important as face-to-face conversations because they represent both the person and the organization. Consequently, special care should be taken to ensure that email and voice mail messages are clear, concise, accurate, courteous, and friendly. When the message is confidential or of a sensitive nature, email and voice mail may not be the best communication methods as neither ensures privacy.

A few general guidelines can be helpful to reinforce effective use of voice mail. It is important to speak slowly and clearly when leaving a voice mail message. Also, it is good practice to give your name, telephone number, and extension number before leaving your message and to repeat them when you are finished. A brief message specifying what the receiver should do—call you back, wait for your return call, etc., should be left. The date and time that you called and the best time for you to receive a return call should be indicated.

When a cell phone is used, courteous behavior should be observed. Cell phone users should keep cell phone conversations brief when in the company of others. Instructors and the learners should turn off cell phones during instruction.

Like voice mail messages, email messages should be brief. The frequency with which email is used for communication makes it prohibitive for busy people to read long messages. Email messages should open with the appropriate salutation, depending on the relationship the sender has with the recipient. Messages should be typed using uppercase and lowercase letters as appropriate. Using all uppercase letters may be interpreted by the receiver as shouting. Correct spelling, grammar, and punctuation should be used. It is best to avoid using abbreviations such as BTW for "by the way," emoticons such as the smiley face, and humor. Messages should be reread and edited before they are sent. Closings such as "Sincerely Yours," or "Very Respectfully," or simply "Thanks" add a nice touch to an email. The email should end with the sender's name, title, telephone number, or other appropriate contact information.

Text messaging (texting) is the common term that refers to using a short message service (SMS). This technology provides instant communication and advantages but, like cell phone use, requires proper etiquette. In the classroom, texting should be avoided. Reading a text message during a conversation, like listening to a voice mail message, is rude behavior. Messages received can be answered at the appropriate time to avoid distracting the class. Texting should never be done while walking or near equipment. In addition, the ability to send text messages should be controlled in an examination setting to eliminate the possibility of cheating.

Text messages are documents that could be accessed by others and require the same care in tone and accuracy as email messages. Slang and inappropriate abbreviations should be avoided. Sensitive information is best communicated in private by a phone conversation.

The ability to communicate effectively carries over from the classroom and laboratory to securing support for the technical training program. Instructors need to work with their industry counterparts and other industry leaders to ensure an up-to-date and viable program. Technical education instructors who form and participate in special industry training advisory councils are better informed of the needs and interests of a particular organization. Instructors who are able to articulate the strengths and needs of their programs clearly and coherently will undoubtedly receive a higher level of support and commitment than those who cannot.

USE APPROPRIATE NOMENCLATURE

Technical and scientific fields of study use specialized terminology or nomenclature. Nomenclature gives special meaning to words when they are used within the context of the specific field of study. Many textbooks on technical subjects include a glossary covering the nomenclature for that subject. For example, "electronics" and "programmable controllers" are examples of nomenclature used for industrial maintenance; "solenoids" and "DC generators" are examples of nomenclature used for electrical motor controls; and "powers" and "roots" are nomenclature used in mathematics. Use of appropriate nomenclature by an instructor is vital to building a learner's knowledge base.

Practice Active Listening

Active listening is as important as speaking clearly. Research shows that most people use only about 25% to 30% of their listening potential. Most of us are distracted constantly by either external or internal noise. Noises generated by machinery, equipment, or coworkers are good examples of external noise. Disinterest in a subject, dislike for the speaker, preoccupation with other responsibilities, or simply being too tired to listen are all examples of internal noise. Most people speak at a rate of 125 to 150 words per minute; most people can listen to approximately 400 to 500 words per minute.

Active listening takes dedication and training. Active listening is paramount for effective teaching. Instructors should practice the following guidelines to improve their listening skills:

- Make and maintain direct eye contact with learners when they are speaking.
- Get involved in the learner's comments by asking questions or acknowledging their comments.

- Do not interrupt the speaker (unless absolutely necessary due to behavior such as improper comments or language).
- Do not change the subject or direct the speaker's comments or questions until you are sure that you understand the intent of the comment or question.
- Keep your emotions under control.
- Respond in a way that lets the speaker know that the speaker's comments or questions were communicated clearly and understood.
- Do not prejudge the learner.
- Empathize with learners who are trying, yet may be having a difficult time grasping the material.
- Ask relevant questions that show that you are listening, interested, and caring.
- Remember: Patience is a virtue!

Perform Administrative Tasks

Record keeping is a vital part of the instructional process. While keeping records may not be a favorite activity of instructors, records are crucial to the operation, maintenance, and credibility of a program. Requests from other instructors and supervisors for information, assistance, or a special report should not be delayed, unless the reason for the delay is appropriate and explained to those who are waiting for the information. Failure to observe requests for certain information may mean a great loss of time and money. Records are necessary, and individual reports are needed in the compilation of final reports and records.

Just as instructors have a set of rules for their classrooms and laboratories, schools have sets of rules that everyone is expected to follow. When a rule appears to be unsound, the instructor should mention the flawed rule to the supervisor along with suggested changes for improvement.

Participate in Professional Development Activities

Professional development in the areas of teaching and learning is just as important as knowledge in a subject area. Although this is one area often overlooked and sometimes discounted by instructors of technical subjects, developing and increasing professional competencies plays a major role in the teaching-learning process. New legislation, state codes and policies, and program guidelines are presented at industry-supported conferences and seminars, as well as state-supported professional development activities.

Universities often offer special courses, seminars, and workshops to help update and upgrade the technical knowledge and professional competencies of instructors, as well as to facilitate the acquisition of new knowledge and competencies. Professional associations provide another source of information through special conferences, meetings, and materials to enhance the professional development of instructors.

Summary

Personal characteristics play an important role in the teaching-learning process. Respect for the learners and their needs, interests, and abilities and enthusiasm for the subject matter are paramount to creating a positive learning environment. Encouragement through providing opportunities for growth and development helps to create situations and conditions in which learners will be self-motivated. It is up to the instructor to provide guidance, direction, and reassurance to learners. All individuals deserve to be treated fairly and equitably. Favoritism creates situations that are counterproductive to learning.

Promptness and attentiveness in the classroom communicate a professional attitude to learners specifically and toward education generally. Successful instructors serve as role models for learners, not only through their actions, but also through their appearance.

Effective communication skills are necessary for success both in and out of the classroom and laboratory. Participants in specialized training programs and workshops will need to communicate effectively to be considered for initial employment and promotions. Good listening skills are imperative.

Teamwork through cooperation helps to ensure successful and productive training programs. Individuals as well as teams of instructors may be responsible for completing administrative tasks. Such tasks should be completed correctly and in a timely manner to avoid possible loss of program funding and to ensure future support for educational programs.

Instructors who stay current in their fields usually earn the respect and admiration of their colleagues, managers, and learners. It is not enough to stay current in the industrial or technical subject areas. Those who want to serve others through the profession of instructing/teaching should also participate in special courses, workshops, conferences, etc., to develop and enhance their instructional technology, professional planning, teaching, monitoring, and evaluation skills.

Introduction

Effective training programs do not just happen; they are the product of careful planning and preparation. The instructional session, whether conducted as a class, a workshop, or a seminar, requires preplanning. In all instructional settings, objectives must be stated, lesson plans or agendas created, and equipment and facilities made ready for use. Also, all the materials should be prepared and software loaded ahead of time. For a workshop or seminar held in an industrial setting, it will be necessary to conduct a needs assessment, and the program will need to be publicized.

Write Objectives for the Instructional Session

Clearly stated objectives provide direction and guidance for the instructional session. Written objectives are a necessary tool in any type of instructional setting and should be written early to allow the instructor enough time to prepare the necessary materials. In addition, clearly stated objectives provide a framework for the learners and for evaluating the results of the instruction. Each objective should include specific conditions, tools, or materials required to achieve the objective. In an industrial setting, objectives should be written after the needs assessment results are analyzed.

Develop a Lesson Plan or Agenda

In a classroom setting, the instructor works from a lesson plan. In an industrial setting, the instructor works from an agenda. The purpose of both is to help the instructor stay on track and stay focused. An agenda should be prepared within two weeks of the workshop or seminar. See Figure 2-1. (For an example of a lesson plan, see Figure 3-2.)

Prepare Instructional Materials

It is the instructor's responsibility to prepare instructional materials. Each instructional session will have its own unique characteristics and goals. The critical question for the instructor is, "How much skill do I expect learners to have and by what date?" Often, the instructor can purchase commercial materials for the session.

A minimum of three to four weeks should be allotted to locate and/or prepare materials. In an industrial setting, vendors may supply materials for training. This is especially true when new equipment is introduced.

SAMPLE AGENDA

Scotchdale Manufacturing Co.
Workshop Agenda–Americans with Disabilities Act
Location: Industrial Site No. 1–Training Room B

Instructor Facilitator: Chris Johnson
Special Guests: Marty Green, State Representative; Bill Fare, Attorney-at-Law

Date: June 15 Time: 8:30 AM

Welcome and Introductions

Overview of Workshop

Overview of Societal Contexts for Individuals with Disabilities

ADA Definitions

Overview of Legal Requirements
- Employer requirements for compliance
- Special situations and specific exclusions
- Nondiscriminatory qualification standards and selection criteria
- Nondiscrimination in the hiring process (recruitment, applications, pre-employment inquiries, and testing)

Health and Safety
- Direct threats to health and safety of self or others
- Direct threat to food handling

Confidentiality
- Post/offer examinations or inquiries
- Limitations on use of medical information

Implications for Scotchdale Manufacturing

Wrap-up–Questions and Comments

Workshop Evaluation

Adjourn

Figure 2-1. In an industrial setting, the instructor uses an agenda to guide the session.

In all cases, it is a good idea for instructors to develop an instructional materials resource file that lists contact information for various sources of instructional material.

PowerPoint® is an excellent software tool for preparing instructional materials. The PowerPoint® program is included in the Microsoft® Windows® software package. It is a user-friendly program that provides a selection of templates for the formatting of slides, sample graphics such as charts and figures, a variety of artwork, an assortment of background colors and designs, and several slide presentation options. It allows flexibility in both use and time by permitting immediate insertion or deletion of information as necessary.

PowerPoint® offers many features that can be used to enhance a presentation. However, care should be taken so that the slide presentation is most effective for the learner. The following general guidelines should be used:

- Use a large enough font size so that the learner can read the slide from any location in the learning environment.
- Include only one major point on each slide, or two minor points, so that there is sufficient white space (background) on the slide to prevent visual clutter.
- Use font sizes and a format that enhance the flow of the information and delineate the main headings and subheadings.
- Use background shading and designs that contrast with the font color so that the text is easy to read.
- Use graphics and artwork that are relevant, attractive, and in good taste.

It is also a good idea to prepare materials that allow for flexibility in use and time. The materials can then be used to meet a variety of current and future needs. The materials should be sequenced in a logical order, with natural breaking points to allow for learner participation. In all cases, instructional materials should be checked for accuracy, completeness, and appearance. Materials should also contain correct content, grammar, and spelling. Instructional materials that are neat, clean, clear, and attractive not only make a favorable impression, they help to stimulate and motivate learning. Such materials speak well of the instructor and communicate to the learners the importance and value of the particular program.

Select and Schedule Equipment and Facilities

Some instructors are fortunate enough to have equipment and facilities dedicated specifically to their program needs. However, in an organization where equipment and facilities are shared among several departments or units, it is critical that the equipment and facilities and any special tools be reserved ahead of time. A two-week notice is usually adequate; however, in some circumstances, such as a particularly busy time of the year, requests for equipment and facilities may need to be made more than two weeks in advance.

The equipment should be checked to ensure it is in safe working condition at least three days before the session begins. Personal protective equipment, such as special clothing, shields, goggles, and gloves, should be available and in good condition. Facilities should be checked to ensure that there are adequate tables and chairs or workstations, that there is enough space for special equipment such as projectors or screens, and that the learning environment is clean and comfortable.

Conduct a Needs Assessment

When an instructional session is held in an industrial setting, a needs assessment should be conducted. It is not unusual to conduct a needs assessment three to six months before instruction. This is particularly true when new products, equipment, processes, or policies are to be introduced. Some employees may need no new training on specific kinds of equipment, others may need minimal training, and others may need extensive training.

The type of training needs and the extent of the training should be determined before any actual planning is undertaken. When feasible, it is a good idea for the instructor to organize a small advisory committee (three to five members) to assist in the clarification and verification of training needs. Management and supervisory personnel as well as wage workers are appropriate members on such a committee.

Sometimes senior workers have accurate knowledge of the kinds of skills and knowledge that individuals need to perform their jobs more effectively. Likewise, management and supervisory personnel may be sensitive to the kinds of training that individuals need. Furthermore, it is not uncommon for industrial trainers to be required to document the type of training needed and to provide a rationale for such training.

The following are the steps for conducting a needs assessment:

1. Select the population to be surveyed. Will information be collected for all employees in the organization, a sample of all employees, or a specific group of employees?
2. Include information on the interests, abilities, and career goals of the target population, as well as their training needs.
3. Set a timetable for completing the survey.
4. Collect the information through survey forms, personal interviews, telephone surveys, or from ideas contributed through the company suggestion box. All of these are acceptable methods for collecting needs assessment information. Document this information in a usable format.
5. Develop the survey instrument using a standardized set of questions. See Figure 2-2.
6. Provide a brief explanation of the survey to the target population. This may be in the form of a cover letter, or it may suffice to include a statement or two on the top of the survey form.
7. Conduct the survey. In the case of written survey forms, the instructor could ask the payroll department to include a form with each person's paycheck. Telephone calls could be made on a schedule so that in the case of shift workers, a representative sample of workers on each shift is contacted. Depending on the information provided, provisions must be in place for submission of survey forms anonymously.

SAMPLE NEEDS ASSESSMENT SURVEY FORM

Employee Interest Survey–Employing the Disabled

Directions: Read each item below and indicate the extent to which you would like to develop knowledge or skills about the item. Base your responses on what you believe would be most beneficial to your current job. Circle the number that corresponds to your response using the response categories provided.

1 = NN = No need/desire to know

2 = SL = Slight importance–would like to know basic information

3 = I = Important–need to acquire knowledge/skills in this area

4 = E = Essential–highest area of need; (Use no more than two to three times.)

KNOWLEDGE/SKILL AREA	Extent of Need			
	NN	SL	I	E
1. Employment practices regulated by Title I of the ADA	1	2	3	4
2. Undue hardship limitations and employers	1	2	3	4
3. Types of reasonable accommodations	1	2	3	4
4. Actions that constitute discrimination	1	2	3	4
5. Health and safety issues and practices	1	2	3	4

Your name: (optional) ____________ Your position: (optional) ____________

Figure 2-2. A survey instrument is used as part of a needs assessment.

It may be necessary to follow up on some members of the target population who do not respond in a timely manner. A follow up may be done in-house simply by reminding prospective participants to return the survey form or via a telephone call.

Analyze Assessment Results

The instructor should have a plan for coding, recording, and analyzing the information after it is collected. A plan for recording and analyzing needs assessment information does not have to be complex. For example, using a blank survey form as a coding sheet, the instructor can tally the number of responses for each item on the survey instrument and then count the tally marks. This method provides a sound basis for determining the number of individuals who need or are interested in a particular kind of workshop or seminar.

Select Participants

After the needs assessment is completed and the results are analyzed, individuals who can benefit from the instruction should be invited to participate in the session. These individuals may express their interest directly or their supervisor may contact the training department. An accurate match between participants' level of ability and interests with workshop objectives is essential for effective and efficient instruction. In some instances, certain select groups must be included in a particular training session, such as when new equipment and procedures are introduced. Likewise, there may be instances when all employees should participate in an instructional program.

Publicize the Instructional Program

In an industrial setting, the instructional program needs to be publicized four to five weeks before it is scheduled to begin. Publicity can take on many forms. For example, announcements included in employees' mailboxes or paycheck envelopes may be appropriate. In addition, special notices might be posted in visible locations throughout the facility.

Summary

In all instructional settings, clearly stated instructional objectives should communicate to learners precisely what they will learn or should be able to do upon completion of the session. A lesson plan or agenda should be prepared to allow learners an opportunity to preview the program.

Instructional materials may be developed by the instructor or they may be obtained from commercial vendors. Accommodations for equipment and facilities should be prearranged with ample time to permit any required operating or safety checks and ensuing repairs. In addition, a plan should be developed to provide appropriate safety gear, as well as for the storage and handling of hazardous materials.

In an industrial setting, a needs assessment should be conducted prior to planning an instructional session. The needs assessment should help clarify the type of training needed, the extent of the training, and potential participants. An advisory committee may be formed to help clarify educational needs for different groups of employees. Participants should be selected on the basis of their needs, interests, and abilities.

Timing is important in advertising a workshop or seminar in an industrial setting. Notices that are issued too far in advance are just as problematic as those issued too close to the workshop date. Time allotted for advertising will vary with the specific type of workshop to be offered, the needs of participants, and organizational policies.

Introduction

One of the most important tasks of the instructor is to plan for instruction. A pre-assessment helps the instructor understand why a learner is in a particular instructional setting and what the learner's expectation is, as well as the learner's ability level. An instructor may need to rewrite materials, organize materials into manageable topics, or develop instructional aids to ensure that all learners benefit from the instruction. A post-assessment lets the instructor and learners know whether the instructional objectives were achieved.

Collect Learner Information

No two individuals are the same. Before meaningful instruction can be planned, the instructor should collect information about the learners. There are several effective and efficient ways in which an instructor can do this. The instructor can engage in discussions with learners or ask them to complete an instructor-developed information form. Questions such as the following help to reveal the learners' interests, needs, and expectations:

- What are your best and worst subjects?
- What are your immediate career plans?
- How do you learn best?
- What type of work experience, if any, have you had?
- How do you spend your spare time?
- Is there any other information that would help facilitate your learning?

Instructor observations of learners are also an effective means of learning about their needs, interests, and abilities. For example, how do learners prefer to spend their free time in the classroom and laboratory? On what types of projects do they prefer to work? Do they prefer to work in a small group or alone? Answers to questions such as these indicate learner needs and interests. Other methods include reviewing school records for related information.

ADDRESS LEARNER COMPUTER SKILLS— DIGITAL NATIVES AND DIGITAL IMMIGRANTS

An instructor can face many instructional challenges resulting from learner differences between digital natives (those who have grown up with digital technology) and digital immigrants (those who have adopted the use of digital technology later in life). Digital natives have had computer experience early in life and commonly view interaction with a computer as a natural activity. This allows them to more easily adjust and navigate to a solution when confronted with software problems and error messages.

In contrast, digital immigrants typically lack this knowledge and confidence. Without a basic comfort level, this makes learning to use and apply software more complicated. This also leads to distraction from the learning process and learner frustration. In these situations, the instructor is forced to divert attention from the class to learners lacking basic skills. Minimizing this interruption to the instructional experience is critical for overall success.

Strategies that can be used to address the divide between digital natives and immigrants include providing prerequisite classes that ensure baseline knowledge and skills, in-class instructional assistants, and instructor-developed and supplied tutorials for the software. Additionally, assessment of the computer skills required for the class and the anticipated computer skills of the learners prior to instruction will help in instructional planning. This is especially important when teaching classes with adult learners and classes utilizing sophisticated software.

Develop Learner-Centered Performance Objectives

An instructor's job is to assist learners in developing the knowledge, skills, and dispositions necessary for them to become productive workers and responsible citizens. Textbooks, instructional manuals, curriculum guides, and course outlines typically state the competencies that learners should be able to demonstrate upon completion of an instructional program. It is important for the instructor to clearly communicate which skills and attitudes are needed to become successful.

Learner-centered performance objectives are the foundation for instructional planning. Performance objectives should include the following:

- a statement of the condition under which the activity should be performed, such as "given a vehicle with worn brake shoes, a technical manual, and proper equipment and facilities"
- an action word or phrase that describes what to do, such as "replace and adjust brake shoes"
- criteria for measuring whether the learner performed the activity successfully, such as "so that brakes function properly"

The time spent writing learner-centered objectives is time well spent. This type of planning facilitates learning and guides the instructional process.

Plan Ahead

Planning the lesson well in advance of the first class gives the instructor time to ensure that the different parts of the lesson are presented in the best sequence and all necessary aids are ready for use. Instructors who spend time getting organized for instruction after the class has begun will find that learners are tardy, inattentive, and not very motivated to learn. Instructors who are hasty may also find that they have failed to include safety instructions, learner exercises, or necessary visual aids.

A good instructor prepares for the instruction by making the necessary copies, accessing media clips and Internet sites, staging handouts, and checking equipment. References in books can be tabbed for quick access to information required. All equipment and instructional resources should be located in the instructional room for maximum efficiency. General tasks for instructional preparation are common to all instructional units. See Figure 3-1.

Most instructors feel a little anxious before the start of a new course or workshop. Being overprepared rather than underprepared helps an instructor overcome anxiety and sets a positive tone for future class sessions. Planning for icebreakers, such as asking learners to introduce themselves or each other, helps learners feel at ease. Likewise, the instructor should not hesitate to share some personal background and special interests with the learners. Such openness lets learners know that the instructor is willing to share information as a participating member of the group.

Another anxiety-reducing technique is to have a small group activity planned for the first class meeting. This not only helps individuals to become acquainted with one another, it is an excellent method for setting the tone for the course or class. The activity may be used to introduce the subject matter as well as to reduce learner anxiety and apprehension.

INSTRUCTIONAL PREPARATION CHECKLIST

ORDER FOR DELIVERY BEFORE INSTRUCTION

- ☐ Textbooks for participants
- ☐ Personal protective equipment, special learner kits, or activity packages
- ☐ Special software

ONE WEEK BEFORE INSTRUCTION

- ☐ Review lesson plan
- ☐ Review appropriate textbook chapters
- ☐ Gather/prepare instructor-provided resources
- ☐ Review supplementary reference materials
- ☐ Review PowerPoint® slides or instructional media
- ☐ Review activity sheets and response keys
- ☐ Duplicate copies of activities for participants
- ☐ Review worksheets and response keys
- ☐ Duplicate copies of worksheets for participants
- ☐ Review participant information sheets
- ☐ Duplicate copies of participant information sheets
- ☐ Review pre- and post-assessments
- ☐ Duplicate copies of pre- and post-assessments
- ☐ Reserve/obtain multimedia equipment
- ☐ Test multimedia equipment for proper operation

DAY OF INSTRUCTION

- ☐ Check condition of instructional facility
- ☐ Test equipment for proper function
- ☐ Cue PowerPoint® slides or instructional media as required

Figure 3-1. A checklist can help instructors prepare more efficiently.

Develop a Course or Instructional Outline

A course or instructional outline presents an overview of the entire course and may be thought of as a teaching plan for the course. A course or instructional outline lists the major subject-matter topics and the sequence of the material to be presented. This allows the instructor to plan the proper pace by which specific course objectives can be fulfilled.

When developing the course outline, consideration should be given to the course objectives, the length of the course, and the instructional facilities available. Material should be presented from the simple to the complex, and new information should build on previous information and provide transfer of similar principles and concepts to different applications. Important components of a course outline or instructional outline include the following:

- name of the course
- broad course objectives
- list of the major units to be presented in the course
- informational topics (discussion, lecture, readings, etc.)
- safety considerations
- demonstrations
- supplementary aids
- learning activities
- evaluation activities

Develop a Lesson Plan

A well-planned lesson, including the acquisition of appropriate teaching aids, equipment, and learning materials, makes the job of teaching much easier for the instructor and of greater value to the learners. A lesson plan should be developed and followed for each classroom or laboratory session. The lesson plan should include the following important teaching information:

- the purpose of the lesson
- learner objectives
- a brief activity or exercise to focus learner attention on the content to be presented
- key elements of content such as specialized terminology, information, concepts, and procedures
- plans for demonstrations of how the information, concepts, or procedures can be applied
- learner activities that encourage learning through multiple senses—visual, aural, and kinesthetic—and provide opportunities for instructor-led and instructor-guided practice
- planned questions to check for learner understanding
- independent learner exercises
- resources needed to teach the lesson
- planned conclusion to the lesson
- a method of assessment and follow-up

Lesson planning is easier when instructors follow a specified format. See Figure 3-2. A predetermined format helps an instructor organize a presentation and assures that all necessary areas of planning have been addressed. In addition, the instructor can know at a glance where to find information on the page each time a lesson is taught. A well thought-out lesson plan reduces all of the classroom instruction to the classic four-step method of teaching—preparation, presentation, application, and evaluation.

The lesson plan format may vary depending upon instructor preference, content to be presented, and objectives to be achieved. Using the same format for each lesson, even when all sections are not necessary for a given lesson, and numbering each section allows the instructor to view each component at a glance, thereby easing the transition from one phase of instruction to the next. The series of lesson plans can be filed in a three-ring binder. Related journal articles, manufacturer technical bulletins, and instructor-created worksheets can be added so that they are readily available. The binder can also be used to store notes, instructional cues, and questions added by the instructor.

Evaluate Instructional Materials

Instructional materials should support the objectives and goals of the instructional program. There are many different types of instructional materials from which to choose. A few examples are textbooks, technical manuals, related instructional media, and special tools and equipment.

LESSON PLAN FORMAT

Exact title of the lesson:

Date:

Session no.:

1. Purpose:

2. Learner objectives:

3. Anticipatory set:

4. Key elements of content:

5. Demonstration:

6. Student learning activities:

7. Planned questions:

8. Independent/group learner exercises:

9. Resources:

10. Planned conclusion:

11. Lesson assessment:

Figure 3-2. A standard format provides ease in following a lesson plan.

All written materials and instructional media such as PowerPoint® Presentations, videos, charts, and props should be previewed prior to their use. Printed materials in particular should be selected carefully to ensure that the reading level is not beyond learner ability levels. There are numerous reliable and easy-to-use readability formulas. These formulas are based on the premise that reading difficulty level is a function of the total number of words in a sentence and the number of multiple syllable words in a sentence. Reading difficulty level is reported as a reading grade level. The Flesch-Kincaid model is available with the Microsoft® Word word-processing program. Another easy to use readability formula is the FORCAST formula, which is especially useful for assessing the readability level of technical materials because of its sensitivity to the large number of multiple syllable words often found in technical manuals and textbooks. See Figure 3-3.

A readability test is only one consideration in the selection process for written materials. While selection of textbooks and technical manuals is somewhat subjective for each instructor, use of a rating scale lends a degree of objectivity to the choice by encouraging consistency in evaluation of materials. Criteria for choosing a textbook or manual may include the following:

- content relevance
- up-to-date content
- cost
- cover design
- style and size of type
- page layout and use of color

- use and appropriateness of graphics
- binding and quality of paper
- table of contents, glossary, index, appendix, bibliography
- unit or chapter summaries
- listing of resource materials
- sequence of content
- review or discussion activities
- laboratory manual
- instructor's guide

FORCAST FORMULA

1. Select a 150-word passage from the technical materials.
2. Count the number of one-syllable words in the passage, for example, 90.
3. Divide that number by 10.

   $\frac{90}{10} = 9$

4. Subtract the answer from 20.

   $20 - 9 = 11$

This yields a readability level of 11, which means that individuals reading below the 11th grade level may have difficulty reading this material.

The formula may be written as follows:

$$\text{readability} = 20 - \frac{\text{number of one syllable words}}{10}$$

Figure 3-3. The FORCAST formula can be used to assess the readability level.

Develop Instruction Sheets for Exercises and Activities

Exercises and activities developed and supplied by the instructor can be used with those obtained from commercial sources. Instructor-developed exercises and activities offer an opportunity for the instructor to apply specific concepts of the instructional unit. The course format determines the amount of time available for activities. Written instruction sheets are helpful and should be based on an analysis of the task to be performed or the knowledge to be acquired and the course objectives. Following is a list of various kinds of instructor-developed instruction sheets that help to facilitate learning.

- Operation sheet—useful for performance-based outcomes. These sheets provide step-by-step instructions for performing a specific skill or task.
- Information sheet—efficient method for providing supplementary information not covered adequately in available textbooks. Information sheets are best used for knowledge-based outcomes.
- Job sheet—valuable aid for use in the early stages of the instructional program. Job sheets list actions (performances) in their proper sequence with sufficient detail to permit performance by the learner.
- Assignment sheet—practical method for making individual and group assignments on secondary reading, creative problem solving, and responding to instructor-planned questions. Assignment sheets may be used as performance tests.

ORGANIZE AND UPDATE INSTRUCTIONAL MATERIALS

Instructors who use their time, energy, and resources wisely know that organization is the key to efficiency. Subject matter should be organized in lesson plans in terms of concepts, principles, and processes consistent with learner backgrounds and readiness to learn new information. Advance organizers such as instructor-prepared study guides for learners and focused discussion questions encourage active participation and help to introduce and integrate new material and previously learned material.

Lesson plans organized in a three-ring binder can be easily removed for teaching, updating, and replacement as necessary. Likewise, special instructional aids, props, models, etc., should be stored in their proper place for safekeeping and easy access. Each time a particular lesson is taught, the instructor should review and evaluate materials and other resources based on changes in technology, learner needs, course objectives, and time limitations.

SUMMARY

Effective instructors develop instructional events, and select and organize materials so that all learners have access to the instructional information. Each learner brings a unique experience, background, and talents to the learning situation. Conducting a pre-assessment of learner knowledge level, along with conducting a survey to collect personal information, is useful for building a profile of each learner.

Such information collected before instruction helps to identify the characteristics of each learner and provides valuable planning information to the instructor.

Instructor-developed exercises and activities, in the form of instruction sheets, allow learners to progress at their own rate by providing for individual differences. The instructor should explain how activities relate to the unit topic and demonstrate each activity step-by-step when appropriate. A course outline provides a comprehensive overview of the sequence of instructional units for the entire course. A detailed lesson plan written in a standard format provides a useful guide for daily lessons.

Lessons should include enough material to allow for flexibility if the lesson proceeds more quickly or slowly than planned. Learner capabilities must also be addressed. The instructor may have advanced learners moving on to more complex activities, so attention can be directed to learners needing more assistance. In some instances, course content is based on selected units. If content areas are covered out of sequence, the instructor must present prerequisite information. Lesson plans are important in ensuring that critical information is covered in the course, workshop, or seminar.

Introduction

A good instructor blends several instructional resources to present information and maximize comprehension. A variety of instructional resources can serve to maintain interest during long sessions. Like methods of instruction, the availability of instructional resources may be dependent on the facility, equipment, and/or budget.

Prepare Textbooks and Other Reference Materials

Reference material includes textbooks, periodicals, instructional manuals, trade journals, technical service bulletins, and company newsletters. Textbooks provide a broad base of reference information for the instructor. User's manuals and articles in technical journals can also be used as references throughout the course. In addition, many companies welcome the opportunity to have their company information distributed to potential customers. Industry and standards organizations can also be contacted for specific information.

The instructor should review references well in advance of instruction. Specific pages may be tabbed in books and manuals for future reference. Information should be presented in the context of the subject being taught.

TAKE ADVANTAGE OF INSTRUCTOR'S GUIDES

An instructor's guide may come in the form of a supplement to a textbook. An instructor's guide often provides some or all of the following kinds of materials: organized step-by-step procedures, solutions to problems, test questions, overhead transparency masters, PowerPoint® presentations, and additional references. In addition, suggestions for developing, organizing, and implementing instructional materials may be included in the instructor's guide.

For the new instructor, the instructor's guide provides a starting point for instruction. As with any profession, the new instructor may feel overwhelmed with the responsibilities associated with a quality instructional program. Experienced instructors can benefit as well from the variety of instructional options and teaching suggestions in an instructor's guide.

USE WORKBOOKS AND WORKSHEETS

Workbooks are designed for use with specific textbooks or learner instructional kits for individual study or as part of the instructional program. Activities apply specific operations and principles taught in the lesson or that require use of the components included in the learner's instructional kit. Applications offer practical examples of key concepts. References to textbooks and required components should be listed for each activity or application. Solutions and answers for all or some selected activities and applications are usually included in the back of workbooks or in a separate answer key.

Worksheets may serve as a catalyst for discussion of content covered in the unit, a review of content covered, a quick quiz, or preparation for a posttest. Worksheets often accompany commercially prepared material or may be developed by the instructor. Like activities, worksheets must be reproduced prior to instruction.

Depending on course needs, worksheets may be handed out to learners at the appropriate time or transparencies may be made of the worksheets and used as an outline for class discussions led by the instructor. Worksheet answer keys should be obtained or developed to list solutions to problems. Copies made from the worksheet masters may be made on colored paper to differentiate instructional components. The instructor should check copyright restrictions before duplicating and distributing commercially prepared material.

PROVIDE COMPUTER ACTIVITIES TO SUPPLEMENT LEARNING

Computers can be useful teaching aids when they are used appropriately. Educational software provides activities for the learner and varies in complexity from simple practice programs to comprehensive interactive simulations. Software designed to keep records such as attendance, assignments, and grades is cost efficient and timesaving. Instructional materials such as learner handouts, slides, and charts can be prepared with a professional appearance using some of the basic software packages.

Instructors have become increasingly reliant on computer software to support instruction. Computer software can be expensive and is subject to the license agreement for a specific use. The terms and conditions of the software license agreement are listed in the program and on the package that comes with the software.

USE PRESENTATION SOFTWARE

Presentation software requires a computer and a multimedia projector. With presentation software, the instructor can utilize a collection of text, photographs, illustrations, video clips, audio clips, and other elements when delivering instruction. Navigational functions of some presentation software enable the instructor to select and enlarge images for greater emphasis. In addition, presentations created by the instructor allow photos and video clips of the specific training programs to be shown, which provides greater topical relevance. Different presentation software products are commonly used, such as Adobe® Acrobat® and Microsoft® PowerPoint.®

Take Advantage of the Internet

There are many instructional resources available on the Internet. Several manufacturers now offer on-line or software-driven technical support for troubleshooting.

Government publications are available on the Internet. For example, current OSHA publications are available at the U.S. Government Printing Office (GPO) web site. Product information is also available from many companies. Information about American Technical Publishers, Inc. training products can be obtained at the ATP web site.

Instructors must evaluate Internet sites critically for accuracy, credibility, and relevance. They should verify that subject matter is correct, with industry-specific terminology and safe processes and procedures. Other considerations include checking that the level of content is appropriate for learners. Finally, instructors should choose web sites that are user friendly. See Figure 4-1.

Videos

Videos allow the presentation of information with or without the instructor present. Videos serve as a valuable teaching tool by presenting operations or activities that are difficult to replicate in an instructional setting. For example, troubleshooting equipment is best illustrated by showing footage of a typical application. An additional benefit of videos is that specific sections (clips) can be shown individually and/or repeatedly as required. Specific operations can be videotaped by the instructor.

WEB SITE EVALUATION

Site title: ___________ URL (Address): _______________ Date: ________

Directions: Use the following key and place a check mark (✓) in the blank opposite each item to record your assessment of each web site.

1 = Poor 2 = Average 3 = Above average 4 = Excellent

CONTENT	1	2	3	4
Accurate	____	____	____	____
Timely	____	____	____	____
Objective	____	____	____	____
Organized	____	____	____	____
Relevant	____	____	____	____
LINKS				
Clearly defined	____	____	____	____
Accessible	____	____	____	____
Relevant	____	____	____	____
GRAPHICS, VIDEO, SOUND				
Accurate	____	____	____	____
Relevant	____	____	____	____
Download speed	____	____	____	____
APPEARANCE				
Attractive	____	____	____	____
Clear	____	____	____	____

Figure 4-1. Instructors should evaluate web sites for accuracy, credibility, and relevance.

Understand the Difference between Copyrighted and Public Domain Material

Instructors must use a variety of reference material to support the content presented in the course. This includes the use of reference material that is copyrighted. Copyright is the legal ownership of literary, musical, or artistic work that authorizes the right to reproduce, publish, and/or sell the work. Copyright is indicated on a work by the word copyright or ©, the name of the copyright holder, and the year. Additional statements may be used to further define limitations for reproduction. The phrase "all rights reserved" is commonly used to specify all derivative use of the work.

Instructors must understand and follow federal legislation related to copyrighted material. The Copyright Revision Act of 1976 is the fourth comprehensive revision of the copyright law. Copyright not only protects authors and publishers, but also benefits society from the creative efforts that are protected. Copyright laws vary in different countries. Copyrighted material may be reproduced within the bounds of fair use for instructional purposes. Reproduction of copyrighted material beyond fair use is illegal.

Permission to reproduce copyrighted material must be requested in writing from the copyright holder. Information in the request should include the definition of material to be reproduced, the manner in which material is to be used, and the number of copies to be made.

The copyright holder will assess the request for the impact on the copyright holder and/or author, impact on the market for the material, and the value of the copyrighted material reproduced. The copyright holder has the right to grant or deny permission to reproduce the copyrighted material and to charge a fee for the use of copyrighted materials.

Some reference material is in the public domain. The public domain is a term used to describe "property of the public." Public domain material is free from copyright restrictions. For example, pamphlets, books, or other material published by the United States government is public domain material. The instructor and instructional institution are best served by using careful consideration when reproducing copyrighted material.

DEVELOP A REFERENCE LIST

Instructors should take care to reference primary sources of information from which copyrighted or public domain material is taken. Primary sources are the original documents that were created at the time the actions or events took place.

Primary sources come in many forms, such as letters, video recordings, and artifacts; however, the technical instructor will mostly be concerned with industry standards, government regulations, and applicable codes. In general, the instructor must be sure that the source of the information is complete, accurate, and up-to-date. Local libraries, professional organizations, the Internet, and publishers, such as American Technical Publishers at ATPeResources.com, provide easy access to a host of primary sources related to technical education.

The Internet can be useful for locating specific documents. Using specific rather than broad terms facilitates a more efficient search. For example, in a search for the Americans with Disabilities Act (ADA), typing in "Americans with Disabilities Act" in a search engine will result in more than 1000 references. It will then be necessary to select the most reliable source, which should be the actual legislation in the federal statutes. In this case, an instructor may choose instead to reference the ADA home page, which provides numerous links to specific parts of the act. In general, web sites with a nonbiased, balanced approach to presenting sources are more reliable than personal web sites or sites where the source material is used to persuade readers to a particular point of view.

Instructors should take care to provide complete information about primary sources regardless of where they are located. Complete information is necessary, not only to give credit to the source, but also to permit the instructor and learners to locate the source if needed. Basic information to include in the reference are the title and author of the document or book or the developer of the web site, name of the editor, name of publisher, place, year, volume number, federal code, the URL if applicable, and the date. Instructors may choose to use one of the various style formats or to put the information in an order they deem most useful to the learner.

Obtain or Assemble Instructional Kits

Learner instructional kits include special components, tools, and supplies needed for laboratory exercises in the instructional program.

Learner instructional kits may be available from a commercial vendor, or the instructor may assemble them. The best instruction utilizes hands-on practice on components from the learner's kit and related hands-on activities performed on equipment found in the field.

Learner instructional kits may be reused depending on the instructional program. However, the instructor must ensure that the contents of kits have all components and are safe and operational for each instructional session. The instructor should maintain a spare kit for replacement of damaged or missing components.

DEVELOP TRAINING STATIONS

Training stations allow hands-on activities on live or simulated equipment. Training stations usually permit a variety of tasks in less space than the actual equipment in industry. Instructors can develop some of the best training stations. For example, a security alarm system training station can be constructed with circuit components mounted on a display board. This allows access to all system components that would normally be located in different parts of a building.

ENSURE THAT EQUIPMENT IS IN GOOD WORKING ORDER

Equipment used in the instructional program must be in good working order and representative of that used in industry. The instructor should check equipment for safe and proper operating condition. For example, guards should be in place and fit properly, fuses and batteries should be in good working condition, and tools should be accurately calibrated. Replacement parts must be available when needed. Some parts are expensive, and precautions should be taken to avoid unnecessary wear and tear in order to reduce the number of replacement parts needed.

Stress Safety

Safety information is the most important information conveyed by the instructor. The instructor must identify potential risks and minimize the potential for injury and equipment damage. Some activities involve greater risk than others. For example, measuring AC voltage at a wall outlet has a greater potential for injury than measuring DC voltage in a flashlight battery. Safety knowledge and practices cannot be overemphasized.

Safety hazards vary with different activities. Equipment and tool manufacturers should provide clearly defined safety precautions and procedures to help the instructor reduce hazards that could lead to an accident. Safety hazards also vary depending on the facility used. For example, a seminar in a hotel conference center may pose fewer safety hazards than an instructional setting within the production facility.

The instructor should be aware that tools and equipment from different manufacturers might not have the same safety features. In addition, some testing equipment may be damaged from previous use. Depending on the instructional program, it may be the responsibility of the learner to provide a test tool. The instructor must not assume that test tools brought to the course are in safe operating condition. All equipment, tools, and related accessories must be tested before instruction begins.

The lesson plan should include warnings, safety precautions, and emergency procedures. The instructor can include additional notes as necessary.

Develop a Relationship with Business and Industry

The relationship between technical instruction and practices in business and industry cannot be overemphasized. Most employers are willing and eager to be involved in instructional programs. An active advisory committee should be established to provide program guidance and job opportunities for participants in the instructional course or workshop.

Employers can provide real-life examples for learners through group tours, practical experiences, presentations, and demonstrations. In addition, equipment and materials contributions from business and industry can greatly enhance technical instruction. Individuals who complete instructional courses, workshops, and seminars are often employed in a local business or industry.

Offer Distance Learning Options

Many educational programs offer distance learning options so that courses can be made available to a wider range of learners. Distance learning is convenient and flexible. When site-based attendance is difficult due to geographical limitations or due to responsibilities that limit a learner's participation in traditional instruction, distance learning is particularly beneficial.

Instructors who provide distance learning may select from a variety of options such as the interactive technologies of telephone and audio conferencing; video tools such as slides, films, and real-time moving images; use of computers for e-mail, real-time computer conferencing, and the Internet; and finally, printed materials such as textbooks, worksheets, and syllabuses.

Distance learning can be as effective as face-to-face instruction if proper preparations by instructors and learners are made. Course requirements and lesson objectives should be planned carefully. Learner needs, along with learner access to required technology, should be assessed before any technology is selected. For example, on-line video demonstrations, simulations, and other digital media require high-speed Internet access. It may be helpful to supplement digital content with print media.

Furthermore, chat discussions and email messages are excellent strategies to reinforce learning. Such strategies provide timely feedback on a learner's progress and contribute to positive learning. Also, strategies that promote and support learning where interaction occurs over time (asynchronous communication) such as email, voice mail, and discussion boards supplement instructional activities.

Finally, depending on a variety of administrative and logistical requirements, distance learning may be blended with on-site instruction. For example, courses that include hands-on laboratory exercises may require learners to report periodically to a regional lab to complete assignments. Also, tests may need to be taken at a proctored regional testing site.

SUMMARY

The instructor may select resources from print and digital media to supplement and enhance learner experiences and knowledge bases. Videos, CD-ROMs, learner kits, and printed reference materials such as textbooks, technical manuals, and worksheets provide practice, review, and tests for comprehension of material presented in instructional units. Unlike material in the public domain, the instructor must get written permission to reproduce copyrighted material.

Basic computer proficiency is a prerequisite for success in virtually every occupational area. The instructor should include content and activities in the course that provide exposure to the computer applications. The instructor serves as a role model for utilizing new computer applications in the field.

The Internet is a tremendous resource and instructional tool that can offer additional resources to the learner. The information available is similar to walking into a huge library; the desired information is somewhere in the library but must be located. The instructor should provide specific instruction in Internet usage to minimize time spent searching. In addition, sources of information on the Internet must be verified as reputable.

The instructional program should serve as a model for safe work practices. The instructor must demonstrate all safety practices and procedures to be followed. In addition, it is the responsibility of the instructor to maintain a safe learning environment.

Introduction

Instructional methods vary with the learners and the content areas covered. The best instruction occurs through the use of several instructional methods. In addition, an effective instructor can quickly sense which instructional methods are the most and least effective in a given situation. In some cases, the instructional method may be limited by time, facilities, equipment, and/or budget. This challenges the instructor to be creative when faced with potential constraints.

Attract and Hold Learner Attention

Valuable time is wasted if an instructor attempts to teach without first securing the attention of all learners in the class. Whispering or talking among learners during the time the instructor is talking confuses other learners and interferes with concentration. The instructor needs to create an environment in which learners are motivated to learn. Explaining the importance and relevance of a lesson helps to bring meaning to the information and creates positive attitudes toward learning. It is easier to sustain the attention and interest of learners when they are involved actively in their learning and the instructor stays focused on the lesson.

The instructor should clearly communicate the intended outcomes and expectations of the instructional unit. The instructor should also explain the relevance of the current lesson in context with material presented previously and information yet to be covered. Information about safety procedures and precautions should be emphasized.

Learners must be challenged but not overwhelmed in an instructional program. A pace that maintains a variety of learner activity and adequate challenge is desirable. Breaks during instruction are necessary to maintain learner alertness. The instructor also needs occasional breaks to maintain instructional vitality. Materials that are unrelated to the current lesson should be handed out and explained at the close of the class period.

USE THE FOUR-STEP METHOD OF INSTRUCTION

The four-step method of instruction—planning, preparation, presentation, and evaluation—is a straightforward and effective method for delivering instruction. In the *planning* step, instructors should keep in mind the importance of matching instruction with the learning styles and preferences of the learners and develop a lesson plan and corresponding handouts that are well thought out. This will help ensure that lesson objectives are appropriate and consistent with individual learner needs and abilities.

The *preparation* step involves getting learners ready to learn by providing an orientation to the lesson. This may be accomplished by providing a brief overview of the lesson, asking preplanned questions before the lesson, or assigning a chapter in the textbook or special problems as homework before they are to be discussed. The planning and preparation steps help to guide the presentation step.

For the *presentation* step, the instructor must decide which delivery method will best serve learner needs and accomplish the objectives of the lesson. Lecture, demonstration, group or individualized instruction, or cooperative learning may be used alone or in combination. For example, if the lesson objective is to impart knowledge related to the theory of hydraulics, the lecture-discussion method is an effective delivery mode; whereas the demonstration method is effective for a lesson designed to teach the correct procedure for taking blood pressure.

The manner in which information is presented affects comprehension. The instructor sets the standard for the way information is disseminated and how the class is expected to respond. Information should be conveyed in the most efficient manner. Instructor interest and enthusiasm for the subject is readily apparent to learners.

Key points listed in the instructional outline should serve as a guide to the content and sequence to follow. Textbooks and other references should be available for reference as information is presented.

The final step is *evaluation.* In this step, the instructor and learner should determine whether the learner has acquired the desired competencies.

ENCOURAGE READING, WRITING, AND MATHEMATICS

Reading, writing, and mathematics are as important today as they have ever been. Failure to include opportunities for learners to enhance these fundamental skills puts them at a disadvantage in terms of employment.

There is a perception that learners in technical courses may not be interested in reading, writing, and mathematics. The technical instructor should not accept this perception and expect success in achieving competency in these subjects. Also, learners are more receptive to mastering these subjects when the instructor ties them directly to common applications and the technical content of the program.

The instructor should provide and show learners how to acquire reading material in their occupational areas. This reading material can include trade magazines, newsletters, catalogs, bulletins, and manuals. Advisory committees are an excellent resource in identifying and securing occupational reading material.

Encourage Thinking, Doing, and Learning

The importance of learning to think cannot be overestimated. The technical worker must adapt to rapidly changing conditions and demands. Skill of hand is still necessary and important, but the ability to retrieve, analyze, and use information in the workplace is becoming increasingly important. Thinking is an internal process that may manifest itself in overt and observable behavior, and habits of thinking and doing are acquired by practice.

It is not unusual for learners to come into an instructional program with previously learned incorrect habits. It is much easier to learn correct habits of doing and thinking than to unlearn incorrect habits. Therefore, one of the most difficult, yet important responsibilities of the instructor is to help learners correct these bad habits by substituting good habits. This is one reason, aside from safety concerns, that instructors should talk the class through the steps of a process or procedure during the process or procedure. This technique helps learners to follow and understand processes and procedures.

Thinking as a means of learning is an important concept. Learning requires forming new associations based on previously learned information and prior experiences. For example, when an instructor helps learners understand the relationship between an instructional lecture and a laboratory experiment, learners see how to apply theory to practical situations. Thus the thinking process becomes clearer, and learning is facilitated.

PREPARE LECTURES

Lecture is an instructional method that uses oral presentation of information to a group. Lecture is an efficient method of quickly disseminating information. Information is conveyed in a one-directional mode. The primary advantage of lecture is the ability to impact many learners at a given time.

PREPARE DEMONSTRATIONS

Demonstration is an instructional method that uses a combination of lecture and instructor activity to convey information. Demonstrations provide an opportunity for the instructor to present or perform skills required in the field. Demonstrations teach proper movements as tools, materials, or instruments are manipulated. The demonstration method is particularly useful for conveying hands-on techniques.

The instructor must prepare for the demonstration in advance to reduce the possibility of errors in performance. A camcorder and large screens may be required to permit visual access to the movements being demonstrated. Using the demonstration method generally promotes greater learner interest than lecture.

DESIGN HANDS-ON ACTIVITIES

A hands-on activity is an instructional method that uses physical contact and interaction during a given procedure. Hands-on activities integrate auditory and visual senses with touch to produce greater comprehension. Hands-on activities that use experimentation can produce valuable learning experiences. Expected results can serve to confirm a theory or principle; unexpected results can serve to elicit curiosity and critical thinking about the causes.

The instructor can use hands-on activities to stimulate learning. For example, substituting a defective component or a component with a different rating for a working component to selected members of the class can spur a discussion regarding the cause of the mixed results.

Promote E-Learning

Electronic learning (e-learning), a form of distance education, is on-line instruction. E-learning provides additional opportunities for individuals to access information and practice skills. With e-learning, the instructor can post discussion topics, problems, and assignments and individual learners can access the information at their own convenience. Learners provide their responses and the instructor provides feedback.

On-line chat rooms and special times set aside for individual interaction with the instructor or peer learners are other features of e-learning. Instructors should be involved in determining the parts of their existing instructional programs that can be transformed into an e-learning environment easily and effectively.

E-learning has gained popularity in industrial instructional programs because of its versatility and economic advantages. E-learning requires its own infrastructure consisting of multimedia and web technologies. Companies may develop their own e-learning infrastructures and content, employ a technology professional or consultant to assist the company in establishing on-line instruction, and train the instructor to conduct and manage the instruction.

Have Learners Develop Educational Games

Another way to reinforce learning is through the use of games. Educational games can be motivating, challenging, and rewarding, especially when the instructor requires the learners to develop their own games to reinforce specific content. Having learners develop their own games requires them to use their thinking, planning, and technical skills.

Games encourage learners to learn by guided discovery. Learners are encouraged to work creatively and collaboratively to develop and enhance their own understanding. Learner-developed games can be played during the instructional period to allow everyone to benefit from them. For instructors who prefer commercially prepared games, there is an abundance of affordable commercially prepared video and computer games on the market.

Use Technology to Reinforce Comprehension

Effective instructors use technology to reinforce learner understanding and comprehension. One method that is gaining increased popularity is the development and maintenance of course web sites. A course web site allows learners to access course information and instructional materials via the Internet at their convenience and when the need arises. Initial development of the web site requires time and planning on the part of the instructor. However, after the site is up and running, the instructor need only maintain and update the posted information.

The home page for the web site might include permanent information about the instructor and course, as well as information that may change periodically, such as due dates for projects or quizzes. Links can be provided that allow learners to download course materials such as the course syllabus, study guides, and PowerPoint® slides. One of the greatest benefits of an instructor-developed web site is that the instructor can preview and evaluate the accuracy and timeliness of online articles related to the subject matter before providing hyperlinks to the articles. A course web site is an excellent way to serve those learners whose work schedules or life styles may not permit them to access the information in more traditional ways.

Video clips and DVDs are being used in many marketing situations to promote products and services. They can also be used effectively to supplement instruction. There are many commercially prepared educational video clips and DVDs on the market. These may provide a good starting point for beginning instructors.

If appropriate and relevant clips or DVDs are not available, instructors can produce their own video clips using a camcorder with real-time demonstrations or storyboards. The instructor can select important segments, edit, or add sound and other special effects to create unique and relevant instructional materials. The clips may be as brief as a few seconds or as lengthy as several minutes. Instructors who have VHS and other analog video tapes can preserve the quality of the tapes by converting them to digital format so that they can edit, copy, or transfer the information to a DVD.

HAVE LEARNERS PARTICIPATE IN JOB SHADOWING

Job shadowing is an instructional method in which the learner is assigned to one or more successful employees at a company to observe the employee performing daily job duties. Many companies allow "shadowing" of employees by instructors and learners so that learners can gain firsthand experience. The shadowing time may be for one day, one week, or longer.

By observing one or more employees for an extended period of time, learners can get an idea about specific job duties and learn state-of-the-art procedures. An added benefit of shadowing is the opportunity for learners to ask questions of the experts at the time the questions arise.

BRING IN GUEST SPEAKERS

Guest speakers used in instructional programs offer perspectives that enhance the content presented. Guest speakers can be chosen from manufacturers, local companies, product service facilities, and suppliers. For example, an electrical equipment distributor can provide current information about new products and capabilities.

Take Learners on Industry Tours

Industry tours provide an opportunity to leave the instructional facility and experience related content-area activities on-site. Instructors should be aware that industry tours require additional planning for transportation and coordination of the host visitation site. Some locations may require personal protective equipment such as eye and head protection.

Prior to the tour, the instructor should explain what will be seen and how it relates to the material covered in class. After the tour, the instructor should review and summarize the information presented during the tour. The instructor should be knowledgeable of personal and institutional liabilities and follow all precautions when taking learners on industry tours.

Attend Seminars and Conferences

Technical update seminar programs by recognized companies present new products, tools, and equipment. Conferences offer a wealth of new information from industry professionals and equipment vendors. Whenever appropriate and possible, instructors should participate in technical update seminars and conferences and encourage learners to participate in seminars and conferences as well.

Conference presenters often provide special handouts or software of their presentations. These kinds of resources can be valuable teaching tools for later review and discussion. Other sources of information include reading trade publications, attending classes, and participating in professional organizations. Conferences also provide opportunities to share common problems and solutions with peer professionals in the field.

PROMOTE COOPERATIVE LEARNING

Cooperative learning uses the collective efforts of several learners to acquire new information. This method is particularly useful when there is a broad range of knowledge and experience in the group. Cooperative learning requires interaction between all learners in the group. The members of the group may be preselected to ensure a variety of individual competencies.

PARTICIPATE IN TEAM INSTRUCTION

Team instruction is an instructional method that involves the use of more than one instructor. Team instruction can offer the advantage of combined industrial and instructional experience. This can provide a richer instructional experience for the learners.

Learner expectations must be consistent among the different instructors. Team instruction also allows an instructor to focus on a particular topic. For example, one instructor may be responsible for special features of some equipment, and the other instructor may be responsible for representative applications in the field.

Practice Problem Solving

Problem solving uses an instructor-created problem to elicit logical reasoning to find a solution. Problem solving should be used only with learners that have advanced capabilities. The learners must possess knowledge of basic concepts, principles, and procedures before attempting problem solving. For example, an advanced seminar activity could include common problems contributing to electric motor failure. Knowledge of electrical theory, troubleshooting techniques, and contributing causes is required before determining the ultimate cause of motor failure. Problem-solving activities must be selected carefully to ensure that they are appropriate for all learners.

The first step in problem solving is to identify and clarify the problem. For example, specific tasks to be accomplished or decisions to be made can pose problems. Learners should know how to state the problem in writing and formulate a separate statement for each problem. After the problem has been stated, it should be analyzed to ascertain the relevant facts, circumstances, or conditions surrounding the problem. It is usually necessary to obtain new information about the problem.

Next, possible solutions should be written. Advantages, disadvantages, and direct and indirect consequences of each solution should be listed. The most appropriate and feasible solution or decision should be selected. The solution should be organized into a plan for implementation and then implemented.

Finally, learners should evaluate the solution or decision and decide if the desired results were obtained. If the results are correct, appropriate, and desired, then the facts were correct and adequate, and the thinking was sound. If undesirable outcomes resulted, the problem solving process should be repeated, taking care to clarify the problem statement and gather information that may have been omitted.

USE WORKSHEETS

Have learners complete worksheets. Depending on the purpose of the worksheets and the type of questions, special references, tools, and equipment may be necessary to complete the worksheets. For example, learners might need to use the textbook to find the answers to worksheet questions. Worksheets can also function as quizzes. The instructor should collect the worksheets for documentation, if necessary.

USE QUESTIONS

Questions are an important part of the learning process and help to correct errors, clarify concepts and procedures, expand learners' knowledge, and monitor and assess learner progress. Questions can be derived from each instructional unit and can function as a guide for review as well as indicators of learner comprehension.

The instructor must attract the attention of the class before asking the question or it will be necessary to repeat it, perhaps several times. Habitual repetition of questions or learners' answers for the sake of those who are not paying attention wastes valuable instruction time and may eventually help develop an attitude of indifference on the part of learners and frustration on the part of the instructor.

At times, particularly during night classes or extended seminars, questions may be used to raise the level of alertness in the class. For example, during a demonstration, if a learner appears to lose concentration, a quick question can redirect the focus of attention. During a troubleshooting procedure, a quick question directed to the learner such as "How does the position of the test leads in this example compare to the other circuit demonstrated?" will get the learner's attention and alert other learners to the need to remain attentive.

Questions should be stated clearly and distinctly so that all learners can hear and understand them. If additional time is available, questions can be used to lead into discussion of related content areas.

Employ Good Questioning Techniques

Calling a learner's name before a question is posed often encourages some learners to tune out. The instructor should state the question first, and then call on a learner to answer, so that all members of the group are required to think about the question. It is better to call on a specific learner so that group recitation is discouraged. Learners who do not know the answer can give the appearance of knowing when the class answers in unison. Volunteer answers should be accepted only under the control of the instructor. Questions should be stated to avoid "yes" or "no" responses. Questions should be framed so that learners are required to summarize information or explain why, how, under what conditions, and to what extent.

Questioning learners in rotation is not the best technique. It indicates who is to answer next and relieves the rest of the learners of the need to listen and think about the answer. The instructor may glance at a seating chart or select roll cards arbitrarily to call on learners. The best method is to learn learners' names as soon as possible so that questions can be asked based on their individual ability levels.

Use Visual Aids

Visualization is one of the most effective ways to learn. Real objects, models, and cutaways are the best aids for demonstrations. When these are not available, simulations, videos, slides, or illustrations may be used. Computer-assisted instruction is also popular and readily accessible in many instructional settings and laboratories today.

Instructional aids should be prepared or secured before the class begins to maximize class time and hold learner interest. All chalkboards or other marking boards should be free of illustrations, notes, formulas, and other information left from the previous lesson. Extraneous information or objects that distract the attention of the learners should be removed prior to beginning the class.

Safety precautions should be demonstrated and explained thoroughly. Failure to explain and demonstrate the proper operation of a tool or piece of equipment can lead to breakage and loss to the school or company, as well as loss of time to the learners.

Assign Learner Presentations

Learner presentations compel the learner to acquire a comprehensive understanding of the content area. This is especially true if a question and answer session is included after the presentation. Insightful questions can arise from the group, and the presenter must be prepared to answer each question adequately. Learner presentations also provide opportunities for advanced investigation into a specific content area.

Individualize Instruction

Individualized instruction is designed to meet the specific learning needs and ability level of an individual learner. Few instructional programs have the resources to provide total individualized instruction. However, an instructor can use some prepackaged instructional modules throughout a course to meet the specific needs of learners. For example, a learner with less field experience than the rest of the class may require additional individualized demonstrations of a particular technique.

Assign Homework

Homework can be a valuable component for learning. A reasonable amount of homework reinforces learning through review of previously learned material, helps prepare learners for further learning, and provides a method of assessing learner progress.

Learning should be made as easy as possible. Therefore, the instructor should provide guidance in locating resources to help learners complete homework assignments successfully. Copying material from a book has little value. Every assignment should require learners to think about facts and concepts.

Homework assignments should be explained fully, along with the date when the assignment is due. Learners should be required to turn assignments in on time. Only extenuating circumstances as determined by the instructor should permit learners to deviate from the due date. Any penalty for late assignments should be communicated to learners before the assignment is due. Every assignment should be evaluated and returned by the next class meeting with appropriate feedback for the learners.

UTILIZE BLENDED LEARNING STRATEGIES

Effective instructors utilize blended learning strategies by incorporating a variety of teaching and delivery methods to accommodate different learning styles. Previous sections in this chapter addressed a range of instructional methods from reading, writing, and mathematics, to e-learning, cooperative learning, and homework. Blended learning strategies are often thought of as integrating e-learning with traditional learning methods. For example, combining virtual and physical resources, such as technology- or computer-based learning, face-to-face learning, and individualized instruction, to achieve an instructional goal would be utilizing a blended learning strategy for that particular goal.

When instructors employ multiple approaches to learning, whether or not the methods involve e-learning and computers or hands-on activities, they are utilizing blended learning strategies. It is the combination, or blending, of the different approaches to learning that is significant.

Give Comprehensive Directions to Assignments

Learners need to understand assignments and the instructor's expectations. Learners should be given an opportunity to think about the process of completing an assignment, and they should understand clearly what has been assigned. Job sheets can help to clarify the assignment. Properly prepared instruction sheets can help to explain how to carry out a specific assignment.

Summary

Effective instructors use a variety of instructional methods and supplementary information during instruction. The method used should be determined by the purpose, objectives, and goals of the lesson as well as by the learners' needs and preferred learning style. The instructor can provide personal experience and insight, examples of the ways in which information is used in the field, and information about the application of the subject matter to the workplace.

The amount of time available for a particular subject or lesson may influence the mode of delivery. For example, e-learning may be an attractive option when time is limited. Lessons should include enough material to allow for flexibility if the lesson proceeds more quickly or slowly than planned. Learner capabilities must also be addressed when selecting an instructional method. The instructor may move advanced learners on to more complex activities so attention can be directed to learners needing more assistance. In some instances, course content can be presented through selected individualized instruction modules.

If content areas are covered out of sequence, the instructor must review prerequisite information. If time permits, additional activities can be developed or selected to enhance understanding or provide more hands-on tasks.

Learners will be better prepared to learn if the instructor explains a lesson before assigning it. After the lesson, a review of the important points helps to fix the lesson objectives in the learners' minds. Likewise, the lesson preview and review make it easier for the instructor to teach the current lesson and prepare for the next lesson.

Instructional skill, like any skill, improves with practice. Courses that cover learning theory and instructional methods are offered for instructors who desire formal instruction. In addition, instructors acquire many instructional skills through trial and error and by emulating successful instructors.

Introduction

Well-organized, safe, and positive classrooms and laboratories are prerequisites to effective teaching and learning. Classroom and laboratory situations and challenges differ with each instructor and group of learners. There are, however, some basic strategies for organizing and maintaining the learning environment that help to ensure high levels of learning and learner satisfaction.

Establish Classroom and Laboratory Procedures and Rules

All learners want to know what is expected of them in terms of work and behavior. Expectations, procedures, and rules related to classroom and laboratory behavior should be established at the beginning of each course. Classroom and laboratory procedures are important because they often relate to safety, emergency situations, or general institutional policies. For example, procedures must be established for severe weather conditions, evacuation, or other extraordinary situations.

The procedures for a course or program should be based on safe and competent technical practices used in industry. Good habits formed in instructional programs carry over to the workplace. For example, procedures for cleaning and organizing a laboratory at the end of an instructional session must be clearly defined. The consequences for not following the procedures should be explained. The procedures should be posted and a copy provided to each learner.

Rules should be limited in number. It is more effective to develop a few rules and enforce them consistently, fairly, and honestly than to develop many rules that may be easily forgotten or confused by the instructor and learners. Rules help to create and maintain a positive learning environment. Learners respect an instructor who enforces the rules that they themselves helped develop.

Create an Active and Positive Learning Environment

A positive learning environment is one in which all learners feel free to participate, ask and answer questions, and put forth ideas without being ridiculed or embarrassed by the instructor or other class members. Instructors should avoid staring at learners, especially timid learners who are struggling to make a comment or answer a question. At the same time, instructors should not embarrass a learner who wants to be the "expert" in everything. Embarrassing any learner in front of the group may make all learners hesitant to speak. Instructors can ask pointed questions or direct statements toward learners to focus their attention and interest on the subject matter.

Alleviate Fatigue

Long lessons are usually fatiguing. Fatigue reduces the efficiency and effectiveness of teaching and learning. Long lessons should be broken into two or more shorter ones by presenting individual topics separately and by tying them together with a preview and summary. Typically, lectures should be no more than 20 minutes. As soon as fatigue is detected, steps should be taken to alleviate the conditions causing the fatigue. Such conditions are particularly dangerous in laboratories, as more accidents occur when individuals are tired or drowsy.

Ventilation and temperature play an important part in the physical condition and comfort of both learners and instructor. A room that is too cold causes discomfort and a room that is too warm is not only uncomfortable, it promotes drowsiness.

In rooms where no provision is made for air circulation, instructors should see that windows are open so that sufficient ventilation is possible or that the doors are opened at necessary intervals to allow for air flow. Learners can think more clearly when there is proper ventilation and the temperature is controlled. However, when doors and windows are opened for ventilation, care should be taken to open them no wider than necessary, so that noise and other distractions outside the classroom or laboratory can be avoided.

If boredom or fatigue is suspected, the instructor should check the room temperature and ventilation and then should check to confirm whether the appropriate method is being used to present the material. If the instructor is unable to account for the fatigue of the learners, discussion questions can be directed at learners to get everyone actively involved. If all else fails, learners should be given a short break to stand up and stretch. The enthusiasm of the instructor is often a good remedy for drowsiness and apathy in the classroom.

DEVELOP SEATING CHARTS

It is not always necessary for the instructor to assign seats. Adult learners often seat themselves. However, a seating chart helps the instructor learn individual names, cuts down on conversation, and aids in maintaining order. Seating charts, when necessary, can contribute greatly to the effective and efficient management of a classroom. When small classes are held in large rooms, the instructor's attention must be focused over a wider area. A seating chart may help restrict movement and therefore be advantageous.

Traditional seating arrangements may not always be the most desirable, especially for group work. A seminar style is appropriate for whole group discussion, whereas small cluster seating is appropriate for small group work. No matter what kind of seating arrangement is used, it should be an arrangement that supports the goals and objectives of the lesson or course.

Required seating is generally unnecessary for older adult learners. However, a chart makes it easier for the instructor to associate names with faces, as well as to pair up individuals to work with special equipment (such as at computer stations). Older adults tend to identify their own seating arrangement, and after the first or second class meeting, they usually settle into a permanent place.

Use Learner Progress Charts

The work of learners should be evaluated, and complete and accurate records should be kept. Progress charts are valuable in keeping up-to-date records. Learners should complete assignments on time. When appropriate, a penalty for late assignments may instill the expectation of personal accountability and responsibility. Learners take greater pride in their work when they know they are being evaluated and when they can see a record of their progress. Progress charts provide a quick overview of each individual's performance. Each progress chart should be confidential, solely for the instructor and learner.

Progress charts in different subject areas may be available through the school or through state departments of education. Commercially prepared progress charts are also available. In some cases, instructors may need to develop or customize progress charts for their individual programs. Whether commercially prepared or instructor-developed, a progress chart should include a list of the important subordinate skills the learner needs to master, the level of mastery, the date of the mastery check, and the signature or initials of the evaluator. See Figure 6-1.

SAMPLE PROGRESS CHART

Competency Profile for CAD/CAM Technician

Directions: Evaluate each participant by checking the appropriate number on the rating scale below to indicate participant's degree of competency.

Name of Participant ____________________ Date ________________

Name of Evaluator ____________________

Ratings:

1 = Not Applicable–Skills in this area do not apply to this participant
2 = Limited Skill–Work requires instruction and constant supervision
3 = Moderately Skilled–Works best with some supervision
4 = Skilled–Works independently with speed and accuracy and no supervision

1	2	3	4	CAD/CAM TECHNICIAN
___	___	___	___	1. Read blueprints
___	___	___	___	2. Estimate jobs
___	___	___	___	3. Develop project schedules
___	___	___	___	4. Create prints to ANSI standards
___	___	___	___	5. Program CNC machine

Signature of Participant ________________ Date ________________

Signature of Evaluator ________________ Date ________________

Figure 6-1. Progress charts provide a quick overview of learner performance.

BEGIN AND END CLASS ON TIME

Instructors should start and stop classes on time, and learners should not be kept beyond the scheduled time. Sometimes enthusiastic instructors are inclined to forget that learners have other responsibilities. It is the instructor's responsibility to end class at the scheduled time so that learners can be on time for other classes, jobs, or family obligations. If learners are asking questions that are of interest to only one or a few learners at the end of the class period, the class should be dismissed on time and those who have special questions can remain for the help they need or want.

Maintain a Safe Learning Environment

Technical instructors should be familiar with the federal occupational safety and health legislation that requires safe and healthy working conditions and environments. Instructors should know, practice, and teach the regulations that relate to their specific area of instruction. Safety practices should be integrated throughout the training program, and the learners should be actively involved.

For example, instruction on donning and using appropriate personal protective clothing should be an integral part of any program. Personal protective equipment may save a learner (or worker) from injury or even death. On-the-job training and internships are valuable experiences that enhance a learner's knowledge; however, mock scenarios in a safe and supervised environment can also be effective techniques for emphasizing the causes and consequences of a worker's actions. Instructors should don and use appropriate safety gear and require that learners do so also.

Instructors should check the general physical conditions in and around their classroom and laboratory to ensure that no safety hazards exist. Stairways, aisles, and floors should be clean and free of debris. Laboratory floors and walkways should be properly marked. The classroom, laboratory, toolroom, and storage space should be clean and organized. The instructor should make daily checks of guards, shields, and other personal protective equipment to ensure that all are in proper working condition. In addition, personal protective equipment that is in good working condition should be available for each learner.

Safety instructions for equipment operation and the use of special tools should be integrated into the course instruction. Safety precautions should be posted near every piece of equipment, along with emergency shutdown procedures. Every instructor should have a first aid plan for emergencies, and the learners should understand the plan and how to carry it out. Horseplay and other disturbances that may occur inside the classroom or laboratory create safety hazards and should not be tolerated.

PRACTICE TOOL AND EQUIPMENT SAFETY

The instructor is responsible for the safe working order and operation of all tools and equipment in the classroom and laboratory. Worn-out tools and malfunctioning equipment can lead to an accident and possible lawsuits. Machines must have proper guards and switches. Safety aids such as an eyewash station, shower, master switches, and fire extinguishers should be available where appropriate. Personal protective equipment such as gloves and safety glasses or goggles should be available to learners and all others entering a laboratory.

Instructors are responsible for providing instruction to all learners on the safe use of tools and equipment as well as instruction on using personal safety devices and wearing appropriate laboratory clothing. A good practice is to have learners sign a form confirming that they have received safety instruction, demonstrated their competence, and agreed to follow safety precautions.

Develop and Use Inventory Systems

Instructors are responsible for the inventory and maintenance of supplies, tools, and machines in their classroom or laboratory. A systematic written plan should be developed to facilitate equipment maintenance and ordering. The system could be a card file using 4″ × 6″ cards to allow space for writing. Ordering information can be kept on one side of the card and maintenance records can be kept on the other. Different colored cards can be used for supplies, tools, and equipment.

Alternatively, the system can consist of 8½″ × 11″ pages filed in a loose leaf binder under sections marked "supplies," "tools," and "equipment." Pages within each section would contain information relating to one tool or piece of equipment such as the date a tool was purchased, the date a machine was serviced, equipment specifications, or the quantity of a consumable material ordered.

An efficient inventory system can be easily maintained using a purchased or developed software program for inventory control. Maintaining an inventory system is an excellent assignment for learners.

Summary

The suggestions presented in this chapter can be useful in helping instructors create and maintain a safe, constructive, and productive learning environment. The old adage "The learners do not care how much you know until they know how much you care" is still applicable today.

One of the first responsibilities of the instructor is to establish a positive environment where learners feel secure and know the course and instructor expectations. The learners need to know that the instructor is interested in them as individuals and in their learning. Securing or developing a systematic way to keep records of learner progress will help in documenting learner competencies and instructional needs, and developing an inventory system will save time. Instructors usually have some control over the physical facilities in terms of heating, ventilation, and placement and storage of tools, and attending to safety hazards and plans for responding to emergencies is critical to a well-organized instructional program.

Introduction

Changing demographics in the American population are influencing the way in which instructors teach learners and the way in which employers relate to employees. An understanding of multiculturalism is essential for maximum effectiveness in the learning environment and productivity in the workplace. Also, an increasing number of individuals with disabilities are joining the workforce. More than ever before, the legal implications of serving diverse populations, especially individuals who are eligible for services under the Americans with Disabilities Act, must be recognized and understood.

Identify Diverse Needs of Learners

It is important for instructors to identify learners from special populations before they are admitted into the instructional program.

Identification can be made by instructor review of learner records, employers, or by the learners themselves. Instructors need to know the academic ability, technical aptitudes and interests, physical capabilities, and life skills of all learners. This information will give the instructor a better basis for planning and serving all learners, especially those from special populations so that adequate and appropriate modifications and accommodations can be made.

In addition, with this information, the learning environment can be modified to properly address the needs of the learners, such as adjusting the room lighting, providing sound amplification, and/or changing the size of monitors or type displayed.

BE KNOWLEDGEABLE ABOUT EDUCATION LEGISLATION

As educators respond to the changing needs of learners and employers, the instructional environment, and the workplace, it is imperative that they be knowledgeable about legislation that impacts education. Laws are now in place to protect and serve every special group currently identified. Some of these laws have a direct impact on the instructional environment and the workplace. Instructors who understand special legislative mandates not only protect themselves and their organization, they also are in a better position to serve all learners effectively and efficiently.

Review Section 504 of the Rehabilitation Act of 1973

Section 504 of the Rehabilitation Act of 1973 prohibits discrimination of otherwise qualified individuals with a disability. The law states that these individuals cannot be excluded from participation in or be denied the benefits of any program or activity receiving federal financial assistance based solely on their disability. The Office of Civil Rights enforces Section 504.

The law was designed to protect individuals with disabilities in an instructional setting or work environment. Until the passage of the Americans with Disabilities Act in 1990, this section of the law was applied when individuals had claims of discrimination in the workplace. Unlike the Individuals with Disabilities Education Act passed in 1990, Section 504 does not mandate that each child have an individualized education program.

Review the Individuals with Disabilities Education Act

The Individuals with Disabilities Education Act (IDEA) passed in 1990 and amended in 1997 applies to individuals who are 3 to 21 years of age. This landmark legislation was enacted to ensure a free and appropriate public education in the least restrictive environment. Free and appropriate public education means special education and related services that include specially designed instruction that meets the unique needs of an individual covered under the law. Unlike

Section 504 of the Rehabilitation Act of 1973, the IDEA requires that an individualized education program (IEP) be developed to meet the unique needs of each individual receiving services.

PARTICIPATE IN THE INDIVIDUALIZED EDUCATION PROGRAM PROCESS

Instructors who participate in the individualized education program (IEP) process can help other professional personnel and the learner to better understand the safety precautions, skills, subject matter, and dispositions required for successful completion of the instructional program. Instructors of technical subjects should prepare a check sheet or obtain one that lists the skills required to be successful in the program. Also, a list of career paths showing various jobs for which the learner may be prepared will help with the planning session. Instructors can help the IEP team set realistic career goals for a learner with special education needs.

Implementing the IEP requires good administrative skills, such as maintaining the necessary records showing services provided, critical dates, assessment procedures used, and learner progress. Such records should be kept confidential and in a secure location where they are easily accessible to the instructor.

REVIEW THE AMERICANS WITH DISABILITIES ACT

The Americans with Disabilities Act (ADA) is perhaps the most powerful piece of legislation ever passed to protect individuals with disabilities in educational and work environments. The ADA states that qualified individuals with a disability cannot be discriminated against in the job application process, hiring, advancement, discharge from employment, employee compensation, job training, and other conditions and privileges of employment. The Office of Civil Rights is responsible for enforcing the ADA.

The ADA defines a disability generally as a physical or mental impairment that substantially limits one or more of a person's major life activities, such as seeing, hearing, walking, standing, sitting, or learning. Any individual with a disability must have the disability documented by appropriate professional personnel. For example, an individual who suffers from a heart condition must have the condition documented by appropriate medical personnel in order to receive special accommodations under the ADA.

Inherent in the ADA is the requirement that educators and employers provide reasonable modifications or accommodations for any qualified individual with a disability as defined under the law. Modifications and accommodations should be made on an individual basis to address the specific needs of a particular individual. Reasonable accommodations may include assigning note takers or readers to assist learners; providing written communication in alternative formats such as in large print or on audiotapes; offering modified examination formats such as oral tests rather than written tests; allowing extended time on tests; permitting the use of certain equipment to aid a learner or worker; providing wheelchair accessibility; and giving preferential seating.

It is important that educators and employers understand their obligations under the law so that they may provide practical solutions to individual needs for modifications or accommodations.

Review the No Child Left Behind Act

The No Child Left Behind (NCLB) Act was signed into law in 2002. This law is a reauthorization of the Elementary and Secondary Education Act of 1965 and redefines the federal role in K-12 education. The law is designed to help close the achievement gap between disadvantaged and minority students and their peers. The law emphasizes four basic principles:

- stronger accountability for results
- increased flexibility and local control
- expanded options for parents
- emphasis on effective teaching methods

Assessment results and state progress objectives are required to ensure that all groups of learners reach proficiency within 12 years. States are required to report assessment results and progress objectives by poverty, race, ethnicity, disability, and limited English proficiency to ensure that no group is left behind. The greater flexibility for states, school districts, and schools in managing the federal funding they receive is meant to enhance teacher preparation, promote the use of technology in schools, help develop and implement innovative programs, and help promote safe and drug-free schools. NCLB contains an initiative to ensure that every child can read by the end of the third grade. Consequently, the number of nonreaders graduating from secondary-level schools should decrease significantly.

The law promises positive implications for postsecondary instruction and industry training. For example, resources once used for remedial or developmental skills could be reallocated for advanced skills and job specific competencies.

Identify Appropriate Instructional Strategies

Various teaching styles and instructional strategies are necessary to accommodate each learner's knowledge and ability level, socioeconomic background, needs, interests, and motivation. The instructor should discuss appropriate learning activities and teaching methods, curricular and facilities modifications, scheduling adjustments, class management procedures, and assessment procedures with each learner needing special accommodations and with the IEP team, if appropriate. The learner may be enrolled in the technical education program for a substantial part of the day; thus, such information will alleviate frustration for both the instructor and the learner and save preparation time.

In addition to using instructional strategies based on the needs and abilities of the learners, the requirements of the job should be analyzed. Prior to placing a learner in an instructional environment, the instructor, program coordinator, or employer should know the essential tasks that must be performed on the job and the level of performance that is acceptable. A job analysis will help to ensure appropriate, effective, and efficient instruction. For example, if the essential functions of a job require long periods of standing, sitting, walking, lifting, or other kinds of physical activities, appropriate instructional strategies should be used to include the kinds of physical skills required in the workplace. The physical and social conditions of the work environment should also be considered.

If employees are expected to work in teams, totally alone, or under difficult physical conditions (extreme heat, cold, noise), instructional strategies should take these conditions into account. In addition, the general skills needed for the job, such as ability to read, write, compute, use certain software, solve problems, or make decisions, should be identified and instructional strategies should be used that require learners to demonstrate mastery of these skills.

COLLABORATE WITH OTHERS

Instructors of technical subjects should plan, in collaboration with special education personnel and other instructors, to provide the best learning environment and activities for learners. For example, the instructor can recommend and help select printed materials for a learner with a hearing impairment or audiotapes for a learner with a visual impairment, or the instructor can provide the special education teacher with a list of terms and concepts used in the technical field.

Summary

Whether learners are in a secondary-level, postsecondary-level, or industry-based instructional program, individuals with disabilities are protected under the law. Key legislative mandates impact both public and private instructional programs. Section 504 of the Rehabilitation Act of 1973 states that an otherwise qualified individual cannot be excluded from participation in or be denied the benefits of any program or activity receiving federal financial assistance based solely on the disability. The Individuals with Disabilities Education Act ensures a free and appropriate public education in the least restrictive environment. The ADA protects qualified individuals with a disability so that they cannot be discriminated against in educational and employment settings.

As the number of minority learners continues to grow, the educational setting and the workplace will become more diverse. Instructors need an objective basis for placing learners in programs, providing modifications and accommodations, and evaluating progress. Employers must be objective in recruiting, hiring, and maintaining a viable, competent, and diverse workforce. A sound understanding of the provisions of the laws that address instructor responsibility will help to ensure that special groups are served appropriately, adequately, and effectively.

A careful analysis of job tasks and essential functions will help instructors and employers alike to focus on the job to be done rather than the individual. Consequently, all individuals in the instructional setting or work environment should be able to realize their potentials and contribute to the efficiency and effectiveness of the organization.

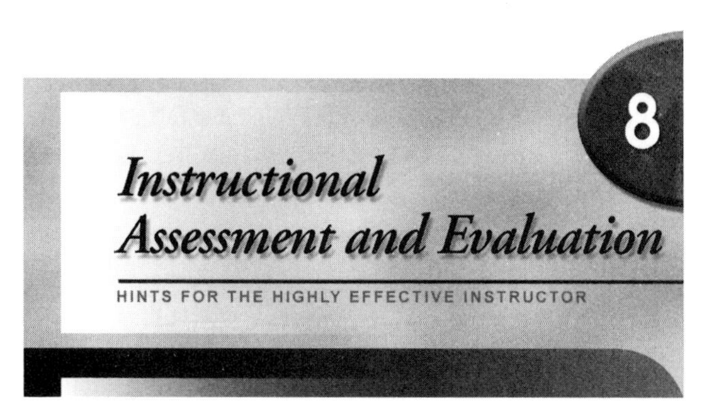

Introduction

Assessment and evaluation are integral parts of the teaching-learning process. Without formal feedback from learners, an instructor can only guess at whether the instruction was successful. Also, accountability for learner achievement is a growing trend in organized educational settings in schools, as well as in industry-based training programs. The rise in accountability and demands of employers for competent workers provide strong encouragement for instructors to collect information to connect learner achievement to instructional practices.

Review Material

Periodic reviews of material covered in each instructional session help the instructor determine whether the learners understand the material. Reviews are also of value to the learners if the review requires learners to react to problems by thinking through solutions and working out applications of the fundamentals of a lesson. Review questions should be stated so that learners are required to apply the information of the pertinent lessons, homework assignments, and work done in laboratory exercises to arrive at correct responses.

The nature of the subject matter should dictate the frequency and content of the review. For example, it is a waste of time to drill on tables and technical information contained in handbooks and manuals used as reference guides for various technical fields. These resources can be made available to the learners. In such situations, it is much more appropriate to review procedures for locating the information, and when and how to use the information, than to review the material itself. It is always appropriate to conduct a brief review at the end of each lesson to provide the preparation step for the next lesson.

UNDERSTAND THE ASSESSMENT AND EVALUATION PROCESS

Assessment is the systematic process of collecting quantitative and qualitative data related to learner achievement. Assessment generally involves testing. Tests provide feedback to the instructor who can discern areas where instruction was effective and areas that need further attention. Evaluation is a judgment that the instructor makes using assessment results to assign scores or grades and give feedback to learners. Assessment, evaluation, and feedback to the learners are the final steps in a course or program.

Assessment serves two different, yet complementary functions. One aim of assessment is to improve learner performance throughout the instructional process by analyzing learners' strengths and weaknesses. This is formative assessment. Formative assessment can be used to improve the design, development, and delivery of instructional courses and programs. Formative assessment should be used throughout the instructional process to make decisions about learner progress and instructional adjustments. Incorrect responses provide evidence of content areas where learners need more assistance.

The other function of assessment is retrospective or summative. This is assessment at the conclusion of a lesson, course, or program to determine the effectiveness of the course in terms of teaching and learning. In summative assessment, the instructor makes decisions related to learner attainment of the skills and knowledge needed to perform certain tasks. In other words, summative assessment can verify the efficiency and effectiveness of a course or entire instructional program.

Development Assessments Based on Performance Objectives

The primary question instructors should ask is, "Do my test items address the objectives of the lesson or course?" These predetermined objectives must be communicated to learners in the form of performance objectives so that learners will know what is expected of them, under what conditions, and to what extent of proficiency. If no objectives have been stated, it is impossible to prepare a valid test because valid test items are those that test against predetermined objectives. It is not feasible to test on all the content covered in a lesson or course. A good test includes a representative sample of relevant lesson or course content.

USE A TABLE OF TEST SPECIFICATIONS

Instructors should analyze the quality of the test items and assessment instruments they use. The first step in developing quality test items is to identify the knowledge and skills to be learned and the levels at which learners are expected to perform. These are the performance objectives. For example, a unit on electrical circuits requires learners to understand safety and electrical circuits, operation of electrical circuits, diagramming electrical circuits, and troubleshooting electrical circuits. The number of test items will depend on the following:

- amount of content to be tested
- depth of the content to be tested
- complexity of the test items
- amount of time allotted for learners to complete the assessment

The format of the items and the emphasis given to each topic within a unit should be based on the amount of instructional time spent on each topic. For example, if considerably more instructional time was spent on diagramming electrical circuits, learners will expect a greater number of test items on diagramming electrical circuits than on the other topics in the unit. Without consideration for the amount of instructional time devoted to specific subject matter, the content taught may be disproportionately represented in the test.

A table of test specifications serves as a useful framework from which relevant and meaningful test items can be developed. The instructor lists the skills and knowledge to be acquired on the left margin with the type of test items to be used listed across the top of the page from left to right. The last column on the right margin shows the emphasis to be given to each topic in the assessment. See Figure 8-1.

After identifying the content areas to be tested and the amount of emphasis to be placed on each topic, the next step is to write the test items. The instructor should avoid writing test items that are too general or too specific. Items that are too general may fail to assess learner achievement of the intended content. Very specific items may test for trivial information. In addition, care should be taken to test only one element of content in each item. An item that tests for more than one concept or skill will give misleading results about the concept or skill the learner has acquired. Developing quality test items requires knowledge, skill, practice, and time.

SAMPLE TABLE OF TEST SPECIFICATIONS				
	Facts, Terminology, Concepts*	Problem Solving*	Application and Integration*	Total percent of emphasis*
Safety and electricity	8	5	5	18
Fundamentals of electrical circuits	7	0	0	7
Electrical prints	3	12	5	20
Industrial applications of electrical circuits	4	16	0	20
Service entrances	3	6	5	14
Troubleshooting electrical circuits	3	8	10	21
Total*	28	47	25	100

* in %

Figure 8-1. A table of test specifications can serve as a print for planning assessment based on instructional content.

Choose or Develop Clear and Concise Assessment Instruments

The appearance of an assessment instrument communicates to learners the importance of the subject and the seriousness of the assessment. Tests should have a professional appearance. Correct grammar and punctuation should be used. The test should be clean and neat. The format of the assessment instrument should be easy to read, with enough white space between items to facilitate reading and writing. Furthermore, the reading level should be appropriate for the content and the learners. Assessment instruments should allow enough time for all learners to complete the assessment.

Clear and concise directions should be provided for each section or set of test items. Directions should be as brief as possible and clearly stated. When a set of items is carried over to another page, it is a good idea to repeat the directions on the new page. The test items should be printed on only one side of the page and should be organized around similar content areas. Easier items should be placed early in the test and followed by more difficult items. This builds learner success and confidence. Also, the levels of responses should vary to include activities that require learners to collect, process, and apply information, as well as demonstrate mastery of facts and figures.

Use Traditional Paper and Pencil Assessments

Traditional paper and pencil assessments are useful for both formative and summative evaluation. These assessments require the learner to select or provide a response. Tests may be computerized so that the learner completes the assessment with either interactive computerized programs or by keyboarding the correct response.

The format for paper and pencil test items should be varied. Test items of a given type should be grouped together in an orderly fashion so that multiple choice items are clustered in one section, true-false items in another section, etc. See Figure 8-2.

CHARACTERISTICS OF TEST ITEMS . . .			
Type	Format	Advantages	Limitations
Completion	True statement in which one or two key words are replaced by blank spaces in which learner writes correct response	Minimizes learner guessing; good for recall of specific facts, figures, formulas	Measures memory rather than understanding and judgment; difficult to develop items that call for one true response
Essay	A statement, question, or scenario that requires an extended response	Freedom for expression; encourages synthesis and evaluation; little opportunity for guessing; easy to construct	Can limit amount of content tested; time consuming for learners to complete and instructor to score
Matching	Two sets of related information arranged in two separate columns, with descriptors in the first column that match short items in the second column	Allows testing of large amount of content in minimum amount of space; objective; easy to score	Difficult to construct items at comprehension level; often fails to measure higher levels of learning; time-consuming for learners
Multiple choice	Statement stem with one correct answer and three distractors	Broad coverage of content; tests at various levels of learning; minimum writing required; reduces guessing; easy to complete; easy to score	Time-consuming to construct; difficult to write plausible distractors; does not measure ability to organize or express ideas

... CHARACTERISTICS OF TEST ITEMS			
Type	Format	Advantages	Limitations
Short answer	Statement or question that requires learner to respond in phrases, sentences, or short paragraph	Easy to develop; measures written communication skills; indicates understanding of content; minimizes guessing	May measure communication skills rather than knowledge of subject matter; scoring may be subjective; may encourage memorization of facts; may have more than one correct response
True-False	A complete statement that is either true or false	Allows testing of wide range and large amount of information in short period of time; easy to score	Difficult to write; may measure memory instead of comprehension; needs large number of items for high reliability; learners have 50/50 chance of guessing correct answer

Figure 8-2. Test items require a learner to select or provide a response.

Create Answer Keys

Keys for the correct responses should be created before the assessment instrument is administered. This early activity can reveal inconsistencies in item wording and problems with anticipated responses, procedures, or calculations. A master key also alerts the instructor to response patterns due to the organization of items on the assessment instrument.

For example, when using multiple choice items, the instructor should be careful that the correct response choices are random, rather than setting up a pattern. Also, the instructor can write the page numbers of the textbook, references, or other sources that give information on that item on the key itself. Answer keys are often available with commercially prepared testing materials. Before the test is administered, the answer key should be checked for accuracy.

ANALYZE QUALITY OF TEST ITEMS

After the assessment instrument has been administered and scored, the instructor can inspect each test item for its level of difficulty. The level of item difficulty is determined by the percentage of learners who answered the item correctly. For example, if 17 out of 20 learners answered correctly, the difficulty level of the item (17 ÷ 20) is 85%. This represents a fairly easy item. A difficult item may be one for which only 6 of 20 learners answered correctly. The difficulty level (6 ÷ 20) would be 30%. The difficulty index ranges from 0.00% (no one answered the item correctly) to 100% (everyone answered the item correctly). Item difficulty level is appropriate only for those assessments that include objective test categories such as multiple choice, true-false, matching, and completion.

An analysis of each item can also indicate how well a particular item discriminated between the high-scoring and low-scoring learners. The item discrimination index is determined by selecting the 25% to 33% of learners who earned the highest total scores on the assessment instrument and the 25% to 33% of learners who earned the lowest total scores on the instrument. The number of learners in the low scoring group and high scoring group should be the same. The number of learners in the low scoring group who answered the item correctly is subtracted from the number of learners in the high scoring group who answered the same item correctly. This number is then divided by the number in each group (not the total number in both groups combined).

For example, if there is a total of 40 learners in a group, the instructor may select the highest scoring 11 learners and the lowest scoring 11 learners for the analysis. If all 11 learners in the high scoring group answered the item correctly and 3 in the low scoring group answered the same item correctly, the item discrimination index would be 0.73. The calculation for this is as follows: 11 – 3 = 8, and 8 ÷ 11 = 0.73. The 18 learners scoring in the middle range are not included in the calculation.

This high index indicates that the item discriminated quite well in a positive direction. That is, more learners in the high-scoring group got the item correct than learners in the low-scoring group. When more of the low-scoring learners get an item correct than high-scoring learners, a negative discrimination index results. A negative item discrimination index indicates a problem with the test item. The problem may be with the wording of the item or the way in which the content was presented. For a total group of 30 or fewer learners, the instructor should use all scores in the analysis by dividing the total group into two groups.

USE ALTERNATIVE METHODS OF ASSESSMENT

Using only one type of assessment may not yield true results of learner knowledge and skills in a particular content area. A variety of assessment techniques and strategies should be used so that all learners have an opportunity to demonstrate achievement. The type of assessment used will vary with the purpose of the assessment and the kinds of decisions the instructor will make based on assessment outcomes. Alternative methods of assessment emphasize qualitative information related to learner achievement. Alternative assessments are growing in popularity. They not only provide optional assessment opportunities for learners, they also enable the instructor to match the form of assessment with specific objectives. See Figure 8-3.

MAINTAIN CONFIDENTIALITY

The instructor should maintain learner confidentiality in all matters related to assessment and evaluation. Therefore, instructors should not post grades or scores by learner name. Posting grades or scores by individual identification numbers is also discouraged. One option is for the instructor to number each assessment instrument and post grades by that number. The number will be known only by the learner who had that specific test. Learners should not know the scores of their peers. The instructor should be careful when returning papers that have been scored so that the score or grade is hidden from the view of others in the group.

ALTERNATIVE ASSESSMENTS	
Type	Characteristics
Journal	Continuous process of learner entries and instructor responses; used for formative evaluation; generates awareness of learner background, interests, experiences, and concerns; identifies prior knowledge of topic; challenges learner to construct concepts related to lesson content; encourages construction of knowledge as a process; encourages open communication between learner and instructor
Oral	Oral expression of ideas; allows for immediate feedback; assesses speech; assesses different levels of learning; checklists may be used to provide written record of performance
Performance	Application of theory, concepts, and procedures to real-life problems; used for formative and summative evaluation; assesses psychomotor skills; may take place in simulated environment; requires learner demonstration of competency; measures content impossible to assess in other ways
Portfolio	Purposeful collection of learner-produced documents (papers, artwork, prints, plans, etc.); demonstrates learner growth and progress over time; demonstrates learner achievement; assesses complementary skills; encourages active learner participation in assessment process; permits evaluation of both process and end products; may include letters from instructors or supervisors; may include examples of learner-produced electronic media (CD-ROMs, videotapes, etc.)
Product	Learner-produced product that demonstrates mastery of content; demonstrates ability to integrate knowledge, skills, and dispositions

Figure 8-3. Alternative assessment methods are growing in popularity.

Distribute Instructor Evaluations

The purpose of instructor evaluations is to provide the instructor with feedback so that instructional effectiveness may be improved. Evaluations help the instructor better understand the way in which learners perceive the teaching-learning process.

The instructor should look for response patterns in the evaluations. While individual opinions are important, it is the majority of like responses that offer the greatest suggestions for improvement. The participation of learners in the evaluation may involve both formal and informal techniques and procedures. Learners who are assured that their evaluation will remain anonymous will tend to take the evaluation seriously and provide honest and worthwhile input.

Teaching is a complex activity and numerous teaching effectiveness inventories have been developed. Teaching effectiveness instruments should solicit opinions on relevant topics such as the personal and professional attributes of the instructor, instructional procedures, and the classroom environment. Instructors who do not involve learners in the evaluation of their teaching are missing a valuable opportunity to encourage learners to participate actively in the learning process and to accept responsibility for their own learning, as well as to gain insight into the effectiveness of their teaching.

Summary

The primary purpose of assessment is to provide information from which the instructor can answer the question, "Did the learners learn?"

Ongoing assessments reveal gaps in participant learning and identify instructional methods that may help to enhance learner achievement. If assessments are to yield valid and reliable results, learners must be prepared to be tested. Large and small group reviews, sample questions, example problems, and study guides help to prepare learners to meet the lesson, course, or program objectives.

A table of test specifications showing proportion of total instructional time devoted to specific topics helps to ensure that content taught is appropriately and adequately represented on the assessment instrument. After administration and scoring of assessment instruments that include objective items such as multiple choice and true-false items, the instructor can ascertain the usefulness of each item by calculating the difficulty and discrimination indices.

Each form of assessment (pencil and paper and alternative) has its appropriate uses, advantages, and limitations. The type of assessment instruments selected for any given body of knowledge or skill should vary with the purpose of the assessment (formative or summative), the predetermined course objectives, learner needs for accommodations, and resources such as time, materials, equipment, and facilities. Performance and product assessments are more useful than paper and pencil assessments when evaluating the extent to which learners can perform a skill. Methods of assessment should be varied to provide opportunities for all learners to demonstrate their competency.

Professionally prepared assessment instruments communicate the importance and seriousness of the evaluation activity. Observing confidentiality of learner progress and achievement protects the instructor, learners, and the institution or organization. Learners should provide valuable feedback to instructors for course and program improvement through teacher evaluations. The feedback should be taken seriously and to the extent appropriate and feasible, and lesson, course, or program modifications should be made. When viewed as dynamic processes, assessment and evaluation are keys to improving teaching and learning.

Introduction

Possessing knowledge of a subject area is not enough for an instructor. Continued development of professional competencies is necessary for effective instruction. Instructors who fail to keep up with the changing technology and conditions in their technical fields may find that some learners know more about new developments than they do. Failure to use the latest equipment that represents industry results in ineffective learning and a loss of learners' interest in the subject.

Take Responsibility for Professional Development

There are numerous opportunities for professional development. New legislation, state policies, and program guidelines are presented at state-supported professional development activities. Universities often offer special courses, seminars, and workshops to help update and upgrade the technical knowledge and professional competencies of instructors as well as facilitate the acquisition of new knowledge. Professional associations provide special conferences, meetings, and materials to enhance the professional development of instructors.

Seek to Increase Subject Matter Knowledge

Subject matter knowledge is acquired through experience in the field and formal training. Experience in the field is necessary for establishing instructor credibility. Most instructors have a variety of industry experiences but may be lacking in a particular skill area. In addition, as technology advances, the instructor must be proactive in acquiring new knowledge and skills.

Technical update seminar programs by recognized companies can present new products, tools, and equipment to the instructor. Conferences can offer a wealth of new information from industry professionals and equipment vendors. These environments may also present the opportunity for instructors to share common problems and solutions with peers. Other methods of acquiring information include reading trade publications, attending classes, and participating in professional organizations.

Establish Advisory Committees

Local business and industry representatives have expertise that can be an excellent resource for an instructional program. An instructor can use these resources by establishing an advisory committee. An advisory committee can offer support and program direction from an industry perspective. Local business and industry representatives are usually supportive of instructional programs because they provide trained workers to the community.

Summary

Instructors must take responsibility for their own professional development. They can subscribe to professional journals, seek new materials, and network with colleagues to gain information and to share ideas. Participating in conferences and seminars is also a way in which instructors can increase their knowledge.

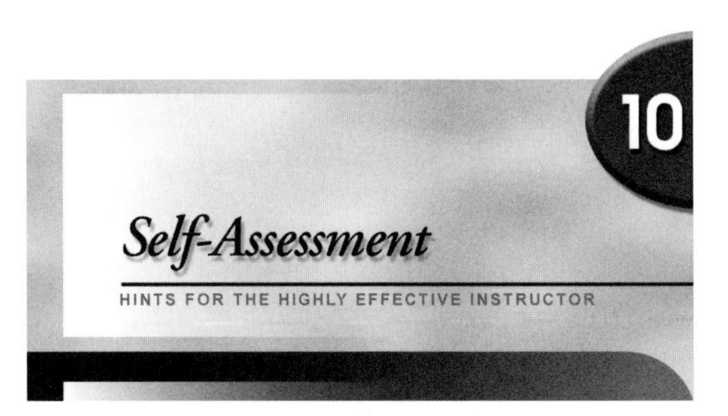

Chapter 1–Instructor Characteristics

Yes	No	Needs Attention	
☐	☐	☐	1. Do you look forward to going to work each day?
☐	☐	☐	2. Could a visitor tell you are the instructor by your attire?
☐	☐	☐	3. Does your attire reflect that expected in industry?
☐	☐	☐	4. Do you encourage your learners to come to class on time or even earlier?
☐	☐	☐	5. Do you encourage your learners to talk with one another and with you before the class is scheduled to start?
☐	☐	☐	6. Are you able to settle differences of opinion and address disciplinary problems without engaging in an argument or confrontation with the learner?

Yes	No	Needs Attention	
☐	☐	☐	7. Are you interested in your learners' progress and growth?
☐	☐	☐	8. Do you feel confident when your supervisor comes to review your class?
☐	☐	☐	9. Are you at ease around your learners?
☐	☐	☐	10. Are you an active listener?
☐	☐	☐	11. Are you comfortable talking to other instructors?
☐	☐	☐	12. Are you comfortable talking to administrators?
☐	☐	☐	13. Do you arrive early to the classroom or laboratory?
☐	☐	☐	14. Are you willing to admit your mistakes?
☐	☐	☐	15. Do you handle differences in opinion diplomatically?
☐	☐	☐	16. Do you look for opportunities to praise your learners when praise is merited?
☐	☐	☐	17. Do you demonstrate the proper etiquette for email communication, cell phone use, and text messaging?
☐	☐	☐	18. Do you have a friendly and cooperative attitude toward your colleagues?
☐	☐	☐	19. Do your learners express a positive opinion about you?
☐	☐	☐	20. Do you perform administrative tasks regularly?
☐	☐	☐	21. Are you supportive of new programs in your school?
☐	☐	☐	22. Do you offer ideas to the administration to improve your program and the school in general?
☐	☐	☐	23. Do you communicate and share materials with other instructors and industrial leaders in your field?
☐	☐	☐	24. Do you interact with the leaders in your field?

Yes	No	Needs Attention	
☐	☐	☐	25. Do you volunteer for additional responsibilities?
☐	☐	☐	26. Do you stay current in your field by taking courses and/or attending seminars?

Chapter 2–Preparation for Instruction

Yes	No	Needs Attention	
☐	☐	☐	1. Do you prepare materials for the instructional session ahead of time?
☐	☐	☐	2. Do you write clear objectives for the instructional session and distribute them to the learners?
☐	☐	☐	3. Do you use a lesson plan or agenda in your session?
☐	☐	☐	4. Do you prepare materials so that you can use them again in the future?
☐	☐	☐	5. Do you ensure that all instructional materials are clear and contain correct grammar, spelling, and punctuation?
☐	☐	☐	6. Do you make sure to schedule all equipment you will need ahead of time?
☐	☐	☐	7. Do you ensure that the equipment you need is in safe and proper operating condition?
☐	☐	☐	8. Do you check the instructional facility ahead of time to ensure that it will meet the needs of the session?
☐	☐	☐	9. Do you conduct needs assessments when necessary?
☐	☐	☐	10. Do you have a method for analyzing the results of any needs assessment conducted?

Yes	No	Needs Attention	
☐	☐	☐	11. Do you select participants, when appropriate, on the basis of their training needs, interests, and abilities?
☐	☐	☐	12. Do you publicize the instructional program?

CHAPTER 3—INSTRUCTIONAL PLANNING

Yes	No	Needs Attention	
☐	☐	☐	1. Do you memorize the names of learners early in the course?
☐	☐	☐	2. Do you make sure that every learner is engaged in productive activity?
☐	☐	☐	3. Do you prepare a written lesson plan for each lesson ahead of time?
☐	☐	☐	4. Are your lesson plans organized and filed for easy access and use?
☐	☐	☐	5. Do you begin your lesson planning with learner-centered objectives?
☐	☐	☐	6. Does each objective include an action word, criteria for measuring the performance, and the conditions for the performance?
☐	☐	☐	7. Do you include appropriate safety precautions in lessons?
☐	☐	☐	8. Do you address learner differences between digital natives and digital immigrants?
☐	☐	☐	9. Do you plan ways to prepare and motivate your learners and develop positive work habits and attitudes in your learners?

Yes	No	Needs Attention	
☐	☐	☐	10. Do your lesson plans include techniques to help learners organize, synthesize, and integrate new information?
☐	☐	☐	11. Do you plan for learner participation, reinforcement, and practice exercises?
☐	☐	☐	12. Do you plan for teamwork?
☐	☐	☐	13. Do you provide learners with feedback?
☐	☐	☐	14. Do you teach learners how to follow directions?
☐	☐	☐	15. Do your lesson plans include a reference section listing instructional materials necessary for the lesson?
☐	☐	☐	16. Do you review each lesson plan for changes and correction before the lesson is presented?
☐	☐	☐	17. Do you organize your classroom and laboratory before the lesson begins?
☐	☐	☐	18. Do you select, prepare, and arrange learner supplementary materials ahead of time?

Chapter 4—Instructional Resources

Yes	No	Needs Attention	
☐	☐	☐	1. Do you use up-to-date technical materials and supplies?
☐	☐	☐	2. Do you use newspaper and magazine articles that present innovations in your technical area?
☐	☐	☐	3. Do you maintain an up-to-date library of books, magazines, and catalogs for your program?

Yes	No	Needs Attention	
☐	☐	☐	4. Do you use an instructor's guide?
☐	☐	☐	5. Do you assign computer activities to learners?
☐	☐	☐	6. Do you use computer programs to manage your instruction?
☐	☐	☐	7. Do you use workbooks and worksheets?
☐	☐	☐	8. Do you use the Internet to find resources?
☐	☐	☐	9. Do you encourage learners to use the Internet to seek out information related to a subject area?
☐	☐	☐	10. Do you use a variety of visual media to supplement instruction?
☐	☐	☐	11. Do you understand copyright law?
☐	☐	☐	12. Have you set up learner training stations?
☐	☐	☐	13. Do you test all equipment to ensure that it is in good working order?
☐	☐	☐	14. Do you repeatedly encourage safety in handling tools and equipment?
☐	☐	☐	15. Do you use transparencies to review key points of your lecture?
☐	☐	☐	16. Do you employ instructional kits to give your learners hands-on practice?
☐	☐	☐	17. Do you ensure that all parts of instructional kits are in safe operating condition?
☐	☐	☐	18. Do you encourage local business and industry to become involved with your instructional program?
☐	☐	☐	19. Do you encourage the development of distance learning programs?

Yes	No	Needs Attention	
☐	☐	☐	20. Are you aware of differences in safety features of tools and equipment from different manufacturers?

CHAPTER 5—INSTRUCTIONAL METHODS

Yes	No	Needs Attention	
☐	☐	☐	1. Do you use a variety of teaching methods?
☐	☐	☐	2. Do you demonstrate the use of tools?
☐	☐	☐	3. Do you use a variety of hands-on activities?
☐	☐	☐	4. Do you invite guests from business and industry to speak?
☐	☐	☐	5. Do you and your learners participate in seminars and conferences?
☐	☐	☐	6. Do you use visual aids?
☐	☐	☐	7. Do you individualize instruction?
☐	☐	☐	8. Do you assign homework?
☐	☐	☐	9. Do you grade and return homework assignments as promptly as possible?
☐	☐	☐	10. Do you discuss homework assignments in class?
☐	☐	☐	11. Do you talk through the processes that you are demonstrating?
☐	☐	☐	12. Do you encourage class discussions among learners?
☐	☐	☐	13. Do you use instructional techniques that enhance learners' critical-thinking and problem-solving skills?

Yes	No	Needs Attention	
☐	☐	☐	14. Do you plan questions for each lesson topic and include these in your lesson plans?
☐	☐	☐	15. Do you ask questions that require learners to describe, explain, or solve a problem?
☐	☐	☐	16. Do you encourage reading?
☐	☐	☐	17. Do you encourage your learners to ask questions?
☐	☐	☐	18. Are your instructional materials up-to-date?
☐	☐	☐	19. Do you use pictures, charts, cutaways, and real objects and equipment in your instruction?
☐	☐	☐	20. Do you prepare and organize materials, tools, and equipment before you conduct a demonstration?

CHAPTER 6—ORGANIZATION OF THE LEARNING ENVIRONMENT

Yes	No	Needs Attention	
☐	☐	☐	1. Is your classroom or laboratory well-lighted, properly heated, and well-ventilated?
☐	☐	☐	2. Are the procedures and rules posted in your classroom and laboratory?
☐	☐	☐	3. Is safety an integral part of your program?
☐	☐	☐	4. Are emergency procedures and safety signs posted in your laboratory?
☐	☐	☐	5. Is your laboratory equipped with first aid equipment such as an eyewash station, shower, and fire extinguisher?

Yes	No	Needs Attention	
☐	☐	☐	6. Do you and your learners wear appropriate personal protective equipment?
☐	☐	☐	7. Is your laboratory equipped with a first aid kit?
☐	☐	☐	8. Do you and your learners know how to administer first aid treatment?
☐	☐	☐	9. Do you and your learners follow the laws on the labeling, storage, and disposal of hazardous waste?
☐	☐	☐	10. Do you and your learners maintain a professional atmosphere in your classroom and laboratory?
☐	☐	☐	11. Do you assign maintenance responsibilities to learners?
☐	☐	☐	12. Is your classroom orderly and neat?
☐	☐	☐	13. Do you maintain an up-to-date inventory of tools and equipment in your laboratory?
☐	☐	☐	14. Do you keep up-to-date maintenance and depreciation records of equipment?
☐	☐	☐	15. Are tools and equipment in your laboratory stored in an orderly and neat manner?
☐	☐	☐	16. Are tools and equipment stored so that they are protected from loss?
☐	☐	☐	17. Are tools and equipment in your laboratory in safe working order?
☐	☐	☐	18. Do you maintain and follow a budget for laboratory tools and equipment?
☐	☐	☐	19. Do you begin and end class on time?
☐	☐	☐	20. Do learners know what is expected of them in terms of procedures and rules?

Yes	No	Needs Attention	
☐	☐	☐	21. Do learners know what is expected of them in terms of work?
☐	☐	☐	22. Do you use the appropriate seating arrangement for the class?

CHAPTER 7–LEGISLATION AND SPECIAL POPULATIONS

Yes	No	Needs Attention	
☐	☐	☐	1. Do you allow for differences in learner abilities?
☐	☐	☐	2. Do you seek out learner records when they are not provided to you?
☐	☐	☐	3. Do you maintain confidentiality of learner records?
☐	☐	☐	4. Do you keep up-to-date on legislative changes?
☐	☐	☐	5. Do you provide challenging and relevant assignments for the gifted learner?
☐	☐	☐	6. Do you use different instructional strategies to accommodate various learning styles?
☐	☐	☐	7. Do you participate in the individualized education program (IEP) process?
☐	☐	☐	8. Have you prepared a list or chart showing the requirements for success in your program to the IEP team members?
☐	☐	☐	9. Do you follow the goals, objectives, and assessment of the IEP for each learner?
☐	☐	☐	10. Do you vary learner assignments and modify the instructional environment to accommodate various learner abilities and interests?

Yes	No	Needs Attention	
☐	☐	☐	11. Do you prepare learners for the enrollment of learners from special populations?
☐	☐	☐	12. Do all learners participate in class discussions and peer interactions?
☐	☐	☐	13. Do you frequently collaborate with special education personnel?
☐	☐	☐	14. Do you confer with learners, parents, counselors, teachers, and other appropriate individuals concerning learner progress?

CHAPTER 8—INSTRUCTIONAL ASSESSMENT AND EVALUATION

Yes	No	Needs Attention	
☐	☐	☐	1. Do you keep up-to-date progress charts for each learner?
☐	☐	☐	2. Do you share progress results with learners?
☐	☐	☐	3. Do you review material before the test?
☐	☐	☐	4. Are written tests clean and neat?
☐	☐	☐	5. Do you write clear and brief descriptions for tests?
☐	☐	☐	6. Do you use a table of test specifications?
☐	☐	☐	7. Do you use alternative methods of assessment?
☐	☐	☐	8. Are test items based on performance objectives?
☐	☐	☐	9. Do tests include a representative sample of material covered in a lesson?
☐	☐	☐	10. Do you grade tests fairly?

Yes	No	Needs Attention	
☐	☐	☐	11. Do you return graded tests in a timely manner?
☐	☐	☐	12. Do you keep learners' grades confidential?
☐	☐	☐	13. Do you use teacher evaluations as a means of acquiring learner feedback for improving your program?
☐	☐	☐	14. Do you conduct follow-up studies to gauge the satisfaction of employers who have hired graduates?

CHAPTER 9—PROFESSIONAL DEVELOPMENT

Yes	No	Needs Attention	
☐	☐	☐	1. Do you read professional education literature?
☐	☐	☐	2. Are you a member of a professional association?
☐	☐	☐	3. Have you chaired a committee or worked actively for a professional association?
☐	☐	☐	4. Have you given a presentation at a professional meeting?
☐	☐	☐	5. Do you help with local school functions?
☐	☐	☐	6. Do you know the supply and demand for employees in your field?
☐	☐	☐	7. Do you visit industries that may hire graduates from your program?
☐	☐	☐	8. Do you maintain an advisory committee for your program?
☐	☐	☐	9. Have you ever written a report or news article about your program?
☐	☐	☐	10. Do you take courses to enhance teaching efficiency?

Bibliography

HINTS FOR THE HIGHLY EFFECTIVE INSTRUCTOR

Emmer, E. T., Evertson, C. M., Clements, B. S., & Worsham, M. E. (1997). *Classroom Management for Secondary Teachers* (4th ed.). Boston: Allyn & Bacon.

Hall, B. H., & Marsh, R. J. (2003). *Legal Issues in Career and Technical Education.* Homewood, IL: American Technical Publishers.

Miller, W. R., & Miller, M. F. (1997). *Handbook for College Teaching.* Sautee, GA: PineCrest Publications.

Miller, W. R., & Miller, M. F. (2009). *Instructors and Their Jobs* (4th ed.). Homewood, IL: American Technical Publishers.

Sarkees-Wircenski, M., & Scott, J. (2003). *Special Populations in Career and Technical Education.* Homewood, IL: American Technical Publishers.

Scott, J., & Sarkees-Wircenski, M. (2008). *Overview of Career and Technical Education* (4th ed.). Homewood, IL: American Technical Publishers.

Storm, G. (1993). *Managing the Occupational Education Laboratory* (2nd ed.). Ann Arbor, MI: Prakken Publications.

ATP
Professional Series

Professional educators are challenged every day as social and economic issues change within a changing population. American Technical Publishers' professional series addresses these important issues and how they pertain to the field of career and technical education.

The series covers a wide variety of topics, from the prevention and proper handling of legal issues to integration of learners from special populations into career and technical education programs. The books in this series serve as references for instructors, counselors, administrators, industry trainers, and human resource specialists.

American Technical Publishers' professional series is designed to help professional educators in the planning and implementation of career and technical education curricula and in dealing with diverse educational challenges. Each book was written by outstanding, knowledgeable authors from the career and technical education profession.

Special Populations in Career and Technical Education

M. Sarkees-Wircenski, J. Scott

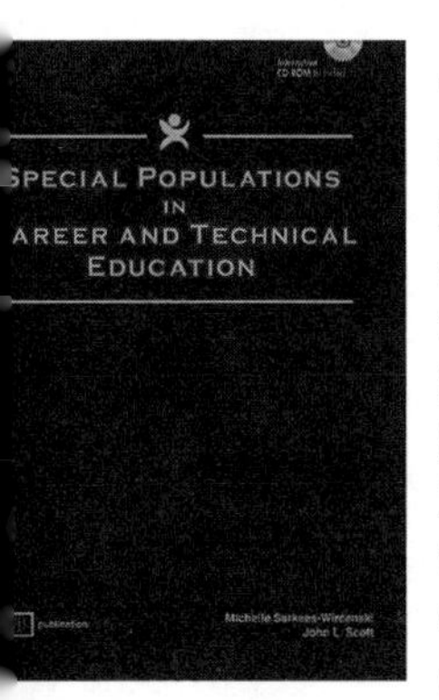

Written for special population instructors, career and technical education instructors, counselors, administrators, and human resource specialists, this textbook covers all aspects of integrating learners from special populations into career and technical programs. Extensive text and graphics identify the characteristics and needs of learners who are disadvantaged or who have disabilities. From the most recent legislation affecting these special populations to comprehensive coverage on assessment and delivery strategies, this reference is the premier source for addressing the unique challenges faced by these learners. This book comes with a CD-ROM.

Highlights:

- Special populations in the workforce
- Curriculum modification
- Individualized education programs
- Student services and organizations
- The transition process

Overview of Career and Technical Education

4TH EDITION

J. Scott, M. Sarkees-Wircenski

This is the preeminent historical reference book in the field of career and technical education. This book takes a comprehensive look at the history of career and technical education, including technical education among diverse populations, and provides in-depth coverage of specific career and technical program areas. It covers the history of legislation affecting education and the workforce in the United States, as well as the philosophies behind past educational movements.

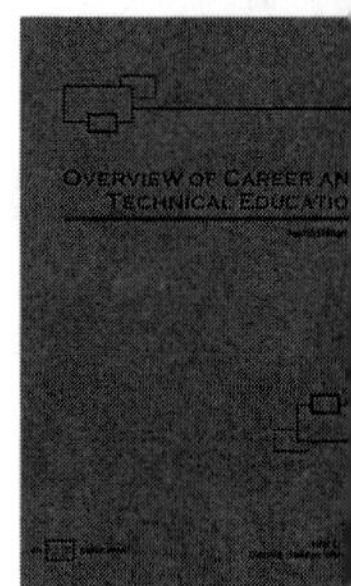

Highlights:

- Defining career and technical education
- Federal workforce legislation
- Development of student organizations
- National student organizations
- Philosophy of career and technical education
- Evolution of career and technical education

Legal Issues in Career and Technical Education

B. H. Hall, R. J. Marsh

Educators are increasingly affected by litigation that can occur from incidents in school settings. This book provides career and technical instructors and administrators with an overview of the legal system and the law and how the law affects career and technical education. This comprehensive reference covers legal research, student organizations, intellectual property, and distance education. It addresses major areas of the law as it applies to career and technical education in both secondary and postsecondary settings. This book comes with a CD-ROM.

Highlights:

- Overview of the law
- Areas of the law
- Federal statutes
- Student issues
- Employment issues
- Hypothetical cases

INSTRUCTIONAL ANALYSIS AND COURSE DEVELOPMENT

2ND EDITION

H. D. Lee, O. W. Nelson

This title serves as an excellent reference for educators developing and implementing instructional courses. Assessment techniques, curriculum design, course development concepts, and program evaluation are covered using concise chapter content and practical case studies. This book integrates theory and application to help guide instructors in building efficient and effective instructional programs.

Highlights:

- Assessment and analysis techniques
- Instructional design
- Instructional development
- Curriculum development
- Program evaluation
- Needs and performance data analysis
- Case studies
- Presentation, report, and proposal development

TECHNOLOGY EDUCATION: FOUNDATIONS AND PERSPECTIVES

Dennis R. Herschbach

Technology Education: Foundations and Perspectives provides a historical overview of various aspects of technology education and the influences that have shaped its evolution over the years. Technology education as a whole is then examined at different instructional levels using examples of innovative programs and issues affecting the future direction of technology education.

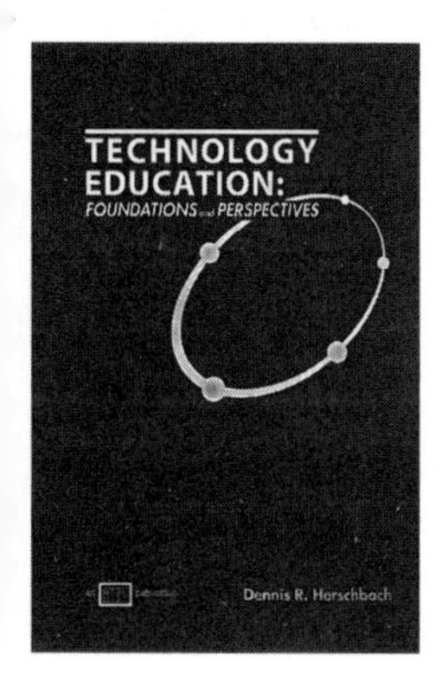

Highlights:

- Beginnings of technology education
- Searching for direction
- Defining technology education
- The instructional representation of technology
- Voices from the field

Writing for Publication

At American Technical Publishers, authors come from many professional backgrounds and possess a variety of professional expertise. However, there is a common bond that links all ATP authors—a commitment to sharing knowledge with instructors and learners in the field. Our authors are usually full-time professionals in addition to being dedicated writers.

The ATP editorial staff possesses a wealth of instructional material development experience. Each staff member has unique qualities that add to the collective capabilities of the Editorial Department. Our technical editors have firsthand experience in industrial programs. Copy editors, illustrators, and support staff have broad publishing and education experience, which ensures the quality and technical accuracy expected in the field. Our mission with all of our products is to develop instructional materials that minimize barriers to comprehension. The editorial staff is committed to this mission.

If you would like to consider becoming an author for ATP, visit our web site at www.go2atp.com and click on the "Writing" section. There you can access a copy of *Authoring Instructional Material.* This document is filled with specific guidelines and examples that show how to organize and develop a proposal that will receive careful consideration.